PHILOSOPHY OF TECHNOLOGY

Other books from Automatic Press ♦ VIP

Formal Philosophy
edited by Vincent F. Hendricks & John Symons
November 2005

Thought$_2$Talk: A Crash Course in Reflection and Expression
by Vincent F. Hendricks
September 2006

Masses of Formal Philosophy
edited by Vincent F. Hendricks & John Symons
October 2006

Political Questions: 5 Questions on Political Philosophy
edited by Morten Ebbe Juul Nielsen
December 2006

Game Theory: 5 Questions
edited by Vincent F. Hendricks & Pelle Guldborg Hansen
March 2007

Philosophy of Mathematics: 5 Questions
edited by Vincent F. Hendricks & Hannes Leitgeb
June 2007

Normative Ethics: 5 Questions
edited by Jesper Ryberg & Thomas S. Petersen
June 2007

Legal Philosophy: 5 Questions
edited by Ian Farrell & Morten Ebbe Juul Nielsen
August 2007

Philosophy of Physics: 5 Questions
edited by Juan Ferret & John Symons
September 2007

PHILOSOPHY OF TECHNOLOGY
5 QUESTIONS

edited by

Jan-Kyrre Berg Olsen

Evan Selinger

Automatic Press ♦ VIP

Automatic Press ♦ VIP

Information on this title: www.philosophytechnology.com

First published 2007

Printed in the United States of America
and the United Kingdom

ISBN-10 87-991013-8-6 paperback

Typeset in $\LaTeX 2_\varepsilon$
Cover photo and graphic design by Vincent F. Hendricks

Contents

Preface

The philosophy of technology is a special region of inquiry. On the one hand, it is continuous with other philosophical topics. For example, practitioners of the philosophy of technology defend their research by appealing to both instrumental and intrinsic justifications that is, they emphasize how their analyses clarify what it means to be human, and portray alternative visions of how humans and non-humans can relate to each other. On the other hand, the philosophy of technology revolves around unique themes and unorthodox methods. A window into these can be found in the following prefatory remarks.

◆

To begin to appreciate why the philosophy of technology has an exceptional status, one only need consider the fact that under current conditions of globalization, technology permeates and connects local and non-local lifeworlds. In public and private spheres, technological forces influence the heterogeneous forms that individual and collective action can take. But even when such influence is powerful and subject to ongoing media attention, it can still occur in subtle ways that are hard for citizens as well as technical professionals to detect. For this reason, philosophers of technology often go beyond merely trying to understand what technology is and how it transforms action, perception, and cognition. In many instances, an activist component is present: visions of the good life are articulated, marginalized voices are represented, and issues of participation and shared governance are explored. In this spirit, one contributor to this volume emphatically declares, "Our contemporary technological culture has many problems probably no more so than many other cultures, but many and I believe that the citizens of a democratic society have the right to expect that technical professionals, including philosophers, will contribute to the solution of these problems."

Moreover, despite widespread claims about the technological character of the present, some contributors argue that it would be a mistake to depict a heightened state of technological affairs as a radical departure from earlier periods of history. As theorists here and elsewhere are now beginning to emphasize, even the seemingly abstract history of ideas cannot be adequately understood without critically assessing underlying material practices and fantasies about material culture. It may even be the case that longstanding notions of intellectual history need to be revisited, if not abandoned altogether. To this end, one contributor goes so far as to suggest, "The model of technical making has shaped the fundamental categories of Western philosophy, existence and essence, nature and culture, and so on." Another contributor insists, "Concepts always develop in embodied relationships with materiality, including the physical arrangement of laboratory space, the potentialities and intractabilities of instruments, and the resistances and enablings that the stuff of the universe offers when humans attempt to manipulate it."

Given the complex nature and broad scope of technological issues, the philosophy of technology has become a thoroughly interdisciplinary enterprise. In fact, only a few professional philosophers actually specialize in the philosophy of technology properly. The majority tend to research dimensions of technology from the vantage point of other specialties, e.g., medical ethics, philosophy of mind, political philosophy, et cetera.

Since many technology scholars come from domains other than philosophy, the philosophy of technology is infused with a vitality and variety of perspectives that perhaps no other branch of philosophy can boast of. This broadness, however, has one pronounced drawback. It is difficult to find an institutional core canon of texts and figures that rigidly defines membership or methodology. By contrast, it seems easier to grasp and track the theoretical trajectories followed by philosophers of science. Divisions such as philosophy of biology or philosophy of physics leave little doubt as to what researchers who occupy these niches attend to.

While the present volume does not aspire to present an uncontroversial canon, it was constructed with the intent of shedding light on two things. First, by including diverse considerations such as neo-Popperian reflections that express skepticism about continental approaches to studying technology, as well as theoretical interventions that question the value of analytic analyses of technology we hope to take leave of the partisanship that typifies all-too-many publications. Our aim, therefore, is to present technology in a pluralistic light that expands upon, rather than reduces, the reasons why it should be considered an exciting domain of inquiry. Second, by highlighting the diversity of theoretical perspectives on technology, we hope to further the impression that one need not be a professional philosopher to think and write philosophically. With this expansive sense of philosophy in mind, we have gathered critical perspectives rooted in anthropology, philosophy, sociology, history, political science, literary studies, and the natural sciences. We have also included contributions that call into question the very stability and adequacy of such traditional disciplinary categories. In this sense, modern, postmodern, and amodern views share center stage alongside perspectives that resist convenient classification.

Ultimately, by taking in the diverse array of reflections, we hope that newcomers and veterans alike will come to better appreciate how complex and exhilarating technological lifeworlds are, have been, and may yet come to be. With this last goal in mind, the contributors were invited to depart from the safety of commenting on completed research, in order to consider the future problems that deserve to be identified and explored. Given the sensitive replies we received, it is our hope that such reflections will not only predict trends, but perhaps inspire future inquiry as well.

Jan-Kyrre Berg Olsen & Evan Selinger
Copenhagen & Rochester
February 2007

Acknowledgements

We are indebted to Elizabeth Pando and Christopher M. Whalin for proof-reading the manuscript; indebted to Claus Festersen for helping us with LaTeX-related problems and grateful to our publisher Automatic Press ♦ $\frac{V}{I}$P, in particular senior publishing editor V.J. Menshy, for taking on this project.

Jan-Kyrre Berg Olsen & Evan Selinger
Copenhagen & Rochester
February 2007

1

Joseph Agassi

Professor Emeritus

Tel-Aviv University, Israel
York University, Toronto, Canada

---

1. Why were you initially drawn to philosophical issues concerning technology?

Quite a few aspects of technology have fascinated me; its inner challenges, its beauty, its great economic, social and political significance, and more. Yet the decisive factor was philosophical, and my study of technology is largely philosophical although also socio-political. The socio-political aspect of it is almost exclusively an attempt to apply Popper's democratic theory to the social problems created by technology: it is the best tool for solving these problems, as it is the theory of democracy and public control, and we need public control over the technocrats. But the main aspect of my work concerns philosophy proper: the two popular philosophies of science, inductivism and instrumentalism, leave no room for the scientific technology that poses a challenge to all great inventors. Inductivists present technology as mere applied science. It is clear that research scientists in industry - chemical, pharmaceutical, aviation, and more - invest much in research that goes well beyond the application of extant science. They also do research in what is known as basic science, namely, they seek scientific theories with the intent to apply them. This, however, is rare; usually they do much more in different directions. Instrumentalists, on the other hand, view science as applied mathematics. This cannot be true, as applied mathematics is a distinct field of research and a fascinating one, but it is neither science nor technology; it is not even applied science. It is different. How? What differentiates applied mathematics, applied science and technology? Technology is autonomous as an intellectual field – although not as a social activity – and as such it is truly fascinating and deserves more attention.

2. What does your work reveal about technology that other academics, citizens, or engineers typically fail to appreciate?

The way technology differs from science. This means that the autonomy of science is more general than that of technology, since technology, though it has a rationale of its own, it is more dependent on social conditions and needs than science is. This enables technologists to appreciate science for its own sake and scientists to appreciate technology for its practical value. Also, it reveals the injustice of the contempt for technology that so many literati exhibit, especially that which is sustained by the irrationalism of their popular anti-scientific philosophers – although, obviously, all contempt is unbecoming anyway. The odd thing is that instrumentalism is presented both as a defense of science and as an expression of contempt for it. Irrationalists view science as mere technology, namely, as devoid of intellectual or cultural value, Hans Georg Gadamer, for example, said that the scientific works of Aristotle interest him more than any text of modern science. In my view this is an excuse for his ignorance, as he will not own it and he will not own that ignorance is poverty and poverty requires remedies, not excuses. The instrumentalists who are pro-science use their instrumentalism as an excuse for not rejecting Newtonian mechanics even though it is false, on the excuse that it is useful. This excuse is silly, since these instrumentalists are often not interested in technology in the least. They will do better to admit that some false theories are of great interest and value - both scientific and technological. The inductivists would likewise learn from this that their view of science as rational belief and of technology as applied science clashes with the applicability of Newtonian mechanics despite its not being a candidate for any belief, rational or otherwise.

3. What, if any, practical and/or social-political obligations follow from studying technology from a philosophical perspective?

The chief corollary of the philosophy of technology is that sociopolitically it is not autonomous: it makes sense only in its sociopolitical context, and it needs social institutions that enhance open-minded social attitudes and thus democracy. Thus, the evils that modern scientific technology brings about invite not the suppression of any technology but the increase of scientific technology and that scientific technology has to include as a significant factor social technologies, and of the character of democratic controls. This should teach us the importance of the rehabilitation of politics by the implementation of the demand from politicians to be more honest than their public, not less.

The current view of technology as morally neutral is sham. The argument that a piece of technology, for example, a handgun, is good or bad depending on its serving the law or the lawbreaker is obviously true; the conclusion that technology is morally neutral does not follow. On the contrary, it follows that any piece of technology is at times good and at times evil. Moreover, the question is not whether or when this or that piece of technology is good or bad. The question is, does the invention of that technology, for example, the handgun, an improvement or not? Does it serve the law more than the lawbreaker? Further, is it true, as we were repeatedly told, that technology will bring unadulterated progress? We once hoped that it will, and now, after Hiroshima, we know that this is not true. And so it behooves us to admit error and not declaim repeatedly that technology is morally neutral. We suggest today that only under some conditions technology will improve human lot. What are these conditions? In my view these are the conditions of effective public control over technology, which is possible only in a modern democracy.

4. If the history of ideas were to be narrated in such a way as to emphasize technological issues, how would that narrative differ from traditional accounts?

The history of ideas should emphasize the difference and the interaction between curiosity for its own sake and curiosity for practical ends. It will not overlook either the difference between curiosity for its own sake, curiosity that is of the investigator into practical matters, and curiosity of the investigator whose work is directed to practical ends. And it will emphasize the value of each of these activities and bring historical examples for each of them. Thus, it will not be biased for or against pure researches, not for or against researches to practical ends in general, and not for or against research for any specific end. It will also emphasize the simple fact that research cannot be evil. The expansion of knowledge is always to the good, even if that knowledge is of how to do evil. But the application of a technology, no matter how benign, can unintentionally bring about some evil consequences; consequently, caution is the right attitude. Each of these points appears in history and can be illustrated by examples that will enrich understanding or science and technology as well as of human affairs in general.

5. With respect to present and future inquiry, how can the most important philosophical problems concerning technology be identified and explored?

The most important question we have now is, what we should do to prevent the extinction of human (and other) life on this planet. This question happens to be philosophical. It belongs to global politics. We should distinguish clearly between international relations and global politics: international relations is as old as nations; global politics is the field of problems which cannot be solved locally but only globally. We cannot solve these overnight, but we can immediately raise hopes for solving them by implementing global programs for their solutions, and these hopes will improve matters at once by reducing the pressures that they cause. These days the migration pressure is on the increase and it will not go away as long as the gap between the poor and the rich countries remains and instills despair in the poor countries of the globe. Efforts to repress this tendency by increasingly powerful technologies, physical and social-political, will only bring about countermeasures from desperate people and so they will defeat them. The only way to reduce this migration pressure is to instill hope. Paul Rosenstein-Rodan has suggested decades ago to institute heavy income tax on the rich nations and on the multi-national conglomerations so as to build infrastructures in poor countries (not for economic aid programs, since these are bound to fail). The idea won no support and was forgotten. A small part of it was recreated by former US President Bill Clinton, who has said, fighting AIDS in Africa rather than in the USA is more useful for the USA and cheaper. Yet he has won no success.

Whenever a program fails like this, the evils that it is meant to fight may go away or be reduced by alternative programs, or the civilization that encounters them may deteriorate to the point of extinction. We have examples of each of these alternatives. We should hope for the best.

Selected Bibliography

Books in English

1. *Towards a Historiography of Science, History and Theory*, Beiheft 2, 1963; facsimile reprint, Middletown: Wesleyan University Press, 1967.

2. *The Continuing Revolution: A History of Physics From The Greeks to Einstein*, New York: McGraw Hill, 1968.

3. *Faraday as a Natural Philosopher*, Chicago, Chicago University Press, 1971.

4. *Science in Flux*, *Boston Studies in the Philosophy of Science*, Dordrecht, Reidel, 28, 1975.

5. (with Yehuda Fried) *Paranoia: A Study in Diagnosis*, *Boston Studies in the Philosophy of Science*, 50, 1976.

6. *Towards a Rational Philosophical Anthropology*, The Hague: Martinus Nijhoff, 1977.

7. *Science and Society: Studies in the Sociology of Science*, *Boston Studies in the Philosophy of Science*, 65, 1981.

8. (with Yehuda Fried) *Psychiatry as Medicine*, Dordrecht: Kluwer, 1983.

9. *Technology: Philosophical and Social Aspects*, *Episteme*, Dordrecht: Kluwer, 1985.

10. *The Gentle Art of Philosophical Polemics: Selected Reviews and Comments*, LaSalle IL: Open Court, 1988.

11. (with Nathaniel Laor) *Diagnosis: Philosophical and Medical Perspectives*, *Episteme*, Dordrecht: Kluwer, 1990.

12. *The Siblinghood of Humanity: Introduction to Philosophy.* Delmar NY: Caravan Press, 1990, 1991.

13. *The Radiation Theory and the Quantum Revolution*, Basel: Birkhäuser, 1993.

14. *A Philosopher's Apprentice: In Karl Popper's Workshop*, *Series in the Philosophy of Karl R. Popper and Critical Rationalism*, Amsterdam and Atlanta GA: Editions Rodopi, 1993.

15. *Liberal Nationalism for Israel: Towards an Israeli National Identity*, Jerusalem and New York: Gefen, 1999. Translation from the Hebrew book of 1984.

16. *Science and Culture, Boston Studies in the Philosophy of Science*, 231, 2003.

Edited books

1. *Psychiatric Diagnosis: Proceedings of an International Interdisciplinary Interschool Symposium*, Bielefeld Universität, 1978, Philadelphia: Balaban Intl. Science Service, 1981.

2. (With Robert S. Cohen), *Scientific Philosophy Today: Essays in Honor of Mario Bunge*, Boston Studies in the Philosophy of Science, 67, 1982.

3. (With I. C. Jarvie), *Rationality: The Critical View*, Dordrecht: Kluwer, 1987.

4. Hebrew Translation of Karl Popper's *The Open Society and Its Enemies*, Jerusalem, Shalem Publications, 2005.

Books by Ernest Gellner, edited by I. C. Jarvie and J. Agassi

1973 Preface to and editing of, *Cause and Meaning in the Social Sciences*, London and Boston: Routledge.

1974 Preface to and editing of, *Contemporary Thought and Politics*, London and Boston: Routledge.

1974 Preface to and editing of, *The Devil in Modern Philosophy*, London and Boston: Routledge. Pp. x - 262.

1979 Preface to and editing of, *Spectacles and Predicaments, Essays on Social Theory*, Cambridge University Press.

1985 Editing and Introduction to, *Relativism and the Social Sciences*, Cambridge University Press.

1987 Editing of, Culture, *Identity and Politics*, Cambridge University Press.

Published over 400 contributions to the learned press.

2

Albert Borgmann

Regents Professor of Philosophy

The University of Montana
USA

---

1. Why were you initially drawn to philosophical issues concerning technology?

Hegel was right. Philosophy should capture its time in concepts (" ... and so philosophy is its time comprehended in thoughts"). "Technology" looked like the most promising vocable to make a start on that project. There was the inspiration of Heidegger, of course, but there was also the encouragement of John Kenneth Galbraith's *The New Industrial State* (1967). Here was an urbane, circumspect, and incisive analysis of the mature technological society. It outlined a revealing system that underlay and comprised seemingly opposed factions, such as management and labor, and seemingly disparate forces, such as the industrial and academic estates. As it turned out, the stress on control and predictability that Galbraith found in the governance of the economy has yielded to the more risky pursuit of affluence; and the convergence of the communist and capitalist economies did not come to pass. But the book has remained a model of engaging social inquiry. It is a reminder as well of how hard it is to comprehend one's time in thoughts.

Technology was also intriguing as a problem of philosophical explanation. There is a theoretical and an empirical side to this. As to theory, the cult of the counterexample in American philosophy is based on the practice of capturing phenomena through necessary and sufficient conditions, and it was unlikely that technology is a phenomenon abstract and jejune enough to be characterized that way. The only alternative seemed to be journalism. The challenge, then, was to uncover a structure that was broadly revealing and conceptually articulate.

The empirical issue comes from the social aspect of technology. An ambitious philosophy of technology has to be ontological, i.e., it has to trace technology in the very fabric of reality. Whatever the texture of that fabric, it surfaces in social institutions and phenomena, and the ways it does so have been carefully gathered and sorted by the social sciences. A philosophy of technology that holds no matter the social science data has to be abstract as a general theory even if it uses concrete instances for illustration. So how is the relation between philosophical theory and empirical evidence to be worked out?

2. What does your work reveal about technology that other academics, citizens, or engineers typically fail to appreciate?

Engineers cling to an instrumental or value-neutral conception of technology. What they design can be used well or ill, they say. People have a half-conscious and rueful awareness that technology is far from neutral and that technological devices and institutions have been coalescing into a force that is hard to resist and not entirely benign. Journalists and essayists are most attuned to the absurdities and frustrations of everyday life in a technological society, but they cannot be expected to have an encompassing theory within which to test and place their observations. Social scientists have comprehensive theories. Those theories are, however, restricted to thin and quantifiable slices of society and culture.

A comprehensive and incisive analysis of contemporary life has to follow up on common intuitions and clarify the ambiguous character of contemporary life. It has to show (1) that technology is an encompassing and structured phenomenon (2) just how it is both beneficial and injurious to human flourishing, and (3) that there is a definite and viable alternative to the rule of technology—models of the good life in a technological society.

Philosophers of technology have provided splendid contributions to the first point, insights more circumspect and original than anything I could have hoped to accomplish. But the normative concerns of the second and third point have generally been treated in a preliminary or anecdotal way. To illustrate, there is to begin with a widely shared intuition that most citizens of the United States are disengaged or passive as far as the shaping of the technological culture is concerned. One response is a call for greater participation. This is a clearly normative and important move. But it is preliminary or penultimate from the moral point of view because it leaves open the question how the culture of technology would be shaped were all voices heard. The implicit assumption that fair procedures will produce morally unquestionable results is surely unwarranted.

Another response is to show that ordinary people have appropriated and reshaped technologies in creative and surprising ways. This too is an instructive and revealing reply. Such a demonstration raises empirical questions, however. What is the percentage of people who are creatively involved in a particular enterprise of technological formation? How many such enterprises are there? How many of the technological circumstances of life do they cover?

Finally, suggestions that technological devices lead to consumption are sometimes met with counter instances of the use of such devices that are engaging and revealing. Claims that the culture of technology is shallow are occasionally countered with evidence of cultural richness. These too are considerable arguments. But how many people are active and probing users of technological devices? And what sort of presence does cultural richness have in the life of a typical citizen in a technological society?

One can dismiss these complaints and argue that the flourishing of the masses is of no concern and that revealing constructive possibilities and instances are philosophy enough. Or one can transpose these cases from the descriptive to the normative key and assert that the cases described are examples of how to live a life of moral and cultural excellence by means of technological artifacts and products. This is a fruitful suggestion, but it is also little more than a starting point, and the normative and empirical issues pop up at a higher level.

The normative proposals need to answer these questions: How do we get from an intuition or suggestion to a moral argument that has a claim on assent? What are the standards of excellence or virtue that come into play here? How does an argument for human flourishing, centered on technology, stack up against the standard kinds of moral argument? And what sort of compelling force can such an argument finally command?

If satisfactory answers to these questions can be found, the empirical question that then arises is this: Assuming that one can embrace technology in culturally and morally admirable ways, what use, as seen from the standpoint of moral or cultural excellence, do people typically make of technology?

Social scientists are trained and equipped to provide valid and reliable answers. The trouble is that they don't ask the question. But there is much indirect evidence about the typical quality of life in the United States that bears on how engaged, informed, creative, and responsible people are in the midst of the technology that pervades their lives. The news is depressing. People are for the most part distressingly ignorant of the world they live in, physically and culturally passive, and indifferent to the miseries of people and the environment.

Surprisingly, perhaps, the empirical evidence takes us around once more to a crucial philosophical question about technology. The dissonance between the constructive and creative possibilities of technology so well described by technological optimists and the distracting and enervating use so widely made of technological possibilities must make one wonder: How come? One reason the question is so rarely raised is a prominent answer that is surely wrong—technological determinism. If one rejects determinism in general, one is left with libertarianism—each one of us is a free and responsible moral agent. But the straightforward application of libertarianism to the technological culture is bizarre: millions of people freely and responsibly make the same deplorable use of the wonderful possibilities technology provides; they make the same unfortunate decisions about eating, exercising, reading, looking, listening, and caring.

The device paradigm explains what it is in the very structure of technological devices that makes the slide from unquestionably beneficial instances to debilitating uses seductive and plausible. It also articulates the locus of responsibility in a realistic way. For most people the blandishments of diversion and disengagement *within* the culture of technology are overwhelmingly attractive. A call for case by case and day after day resistance is unhelpful. But everyone shares responsibility for the culture of technology *as a whole*, and everyone is capable of assuming that responsibility as long as there is some shape and intelligibility to that whole. The device paradigm discloses that structure and how it is to be understood.

Of course the device paradigm is not the only answer to the question as to the coherence and momentum of our society. There are Marxist, capitalist, religious, and feminist proposals among others. This competition of social paradigms can only be adjudicated morally, by showing what the good life finally is. On that basis one has to demonstrate which paradigm most clearly reveals the obstacles and opportunities of the good life. A crucial challenge for this ethically subsidiary demonstration is an analysis of the subversive power of technology that is often and dimly gestured at under the headings of commodification and consumerism. Focal things and practices have the concreteness and robustness to withstand technological subversion and to flourish within a technological society.

The presentation of focal things and practices is a precarious affair. They are easily dismissed or ridiculed. But a normative answer to the moral question of technology needs to have something like their concreteness and definition. Proposing necessary conditions or illustrative cases of the good life puts one in a less vulnerable condition. Such considerations are by their nature more inclusive and appear to be dispassionate and liberal. But consider the technological innovations that at their inception were hailed as conditions conducive to a democratic life of cosmopolitan awareness and cultural energy: newspapers, cars, radio, television, the interstate highway system, the Internet. All of them have in part become utilities that help to maintain society as a going concern. But where they extend beyond means and shade over into final goods, they serve passive consumption. Hence any positive proposal needs to pass this test: Given the history of technological culture and its results, how likely is the proposal to resist adaptation to mindless consumption?

So what is the device paradigm and what are focal things and practices? These are questions I have answered many times. My point was not to answer them once again, but to locate the answers in the common understanding of technology among academics, engineers, and citizens.

3. What, if any, practical and/or social-political obligations follow from studying technology from a philosophical perspective?

The ruling liberal democratic theory tells us that once individuals are provided with the basics of a decent life, no one, least of all the government, is to tell them how to live their lives. Within limits, this principle is morally unimpeachable. But its actual limits are much narrower than the dreams of theory imagine them. Liberal theorists assume that the basics of life are value-neutral. But if there is one thing that the significant philosophers of technology agree on, it is this: Contemporary culture is pervasively technological, and technology is not neutral. Hence technology studies need not impose moral obligations, nor can they avoid them. Studying technology is *per se* involved in moral obligations.

To discharge them adequately, one needs to confront the conundrum of necessary conditions once again. Sustainability, e.g., is a crucial moral condition of human well-being. But if success in securing it is likely nothing more than allowing a majority of consumers to be sustainably indolent, then doing nothing more than making the ethical case for a sustainable way of life is to side-step the crucial moral task.

How, then, do we strike a balance between meddling with people's lives and taking comfort in morally inconclusive proposals? The challenge is to recognize the momentum of the institutional forces that conduce to a life of apathy and to counter that momentum with institutions that favor a life of engagement without encroaching on the unquestionably beneficial and indispensable structures of technology.

The philosophical and moral task gets traction from two terms that are commonly, if imprecisely, used to criticize technology—commodification and consumerism. The device paradigm helps us to give these intuitions precision and context. "Commodification" is used to indict the diminishment things and practices suffer when they are drawn into the market and made available for sale and purchase. But the economic definition fails to capture the moral censure. Not everything is cheapened when commodified, and many commodifications are in fact beneficial. What is needed, then, is a philosophical conception: A thing or a practice gets commodified when it is detached from its context of engagement with a time, a place, and a community. The result is a commodity in a philosophical sense. (The economic and philosophical definitions largely overlap but they do not coincide.) Such detachment occurs when intelligible engagements are replaced by some machinery. Mechanization and commodification go hand in hand. This conjunction is parallel to the machinery-commodity structure of the device paradigm.

A commodity in the philosophical sense is attractive because it is available in the abundance and variety mechanization makes possible and, more important, because each such commodity is available without the burdens of engagement. Taking up a commodity is consumption in a correspondingly philosophical sense that articulates the moral censure implied in "consumerism." In the basic sense of the word, to consume is to stay alive. To consume in the philosophical sense is to ingest commodities, and if this becomes the dominant way of life, passivity and apathy result. That's the moral calamity.

To counter it, public policy needs to favor places, times, and communities of engagement. This is a matter of recognition rather than invention. The promoters of the new urbanism, of farmers markets, of baseball and soccer, of parks and running trails are already taking on the challenge of decommodification or what may be called realization. There is, needless to say, a personal analog to such public endeavors. What needs saying is that realization is not an all-or-nothing affair. Contexts of engagement have a range of scope and thickness, and any broadening and deepening of those contexts is a good thing.

Two more points have to be mentioned to answer plausible skepticism. To complete the moral argument about technology one has to show how virtues and contexts of engagement hang together. And one needs to add that immediate work for social justice and environmental stewardship can't await the invigoration engagement. Progressive policies and real engagement need to be pushed side by side. At length they may intersect this way: The great obstacle to justice and stewardship is not lack of knowledge or resources, but the fusion of ignorance and indifference that has paralyzed the large middle portion of the population. Engagement may break the spell of indolence.

4. If the history of ideas were to be narrated in such a way as to emphasize technological issues, how would that narrative differ from traditional accounts?

Marx described the rise of commodification well and appreciated its epochal force. But in important ways he misunderstood it and drew fallacious inferences from his misinterpretation. What's needed, especially in the case of the United States, is a resumption of Marx's project to show how underneath the crises and conflicts this country has endured, economic and moral commodification was the great project that has lent energy and direction to our history and has now reached the end of its vigor.

5. With respect to present and future inquiry, how can the most important philosophical problems concerning technology be identified and explored?

The Philosophical concern with technology faces an internal and an external challenge. The internal challenge comes from the sociology of the profession. The predicament of contemporary American philosophy is its self-absorption. The more prestigious the philosophy and its practitioners, the more self-absorbed. First-rate analytic philosophy is rigorous, sophisticated, and interesting. But it makes less and less of a difference the further one is removed from the centers of analytic excellence. Professional prestige has not been a problem for philosophy of technology. But as the discipline matures and gets more sophisticated, it may get co-opted by the analytic school as continental philosophy is in danger of being. Bright young philosophers of technology may get prestigious appointment, but with the expectation that their emphasis shift to analytic philosophy of science (albeit technoscience), epistemology, metaphysics, or ethics.

Although such a development is devoutly to be wished for, the danger is that the high analytic style of philosophy will continue to sap the philosophical concern with the concreteness of life and the reform of society. Hence philosophy of technology needs to seek a conversation not just with the philosophical elite, but as much with social scientists, fiction writers, and journalists.

The external challenge comes from a sea change in the standing of technology. Whatever ideological upheavals may have shaken the world in the last century, the progress and blessings of technology have never come into serious popular question. But the entire globe is now being singed by the thoughtless forcing of industrialization. Some of the European countries have become unsure of the promise of technology. Many Muslims have serious reservations. Even in the United States the technological pursuit of happiness is more and more being questioned. At the same time, the gigantic populations of India and China are hotly pursuing technological progress, and billions of sick or starving people urgently need the relief of technological prosperity. To illuminate and clarify this confusing situation and to propose incisive remedies, new philosophical approaches to technology may be needed.

Selected Bibliography

Technology and the Character of Contemporary Life: A Philosophical Inquiry. Chicago: University of Chicago Press, 1984; 5th printing 1997.

Crossing the Postmodern Divide. Chicago: University of Chicago Press, 1992; 5th printing 1998.

Holding On to Reality: The Nature of Information at the Turn of the Millennium. Chicago: University of Chicago Press, 1999; 2nd printing 2000.

Power Failure: Christianity in the Culture of Technology. Grand Rapids, MI: Brazos Press, 2003.

Real American Ethics: Taking Responsibility for Our Country. Chicago: University of Chicago Press, forthcoming.

3

Mario Bunge

Professor of Philosophy

McGill University
Canada

---

1. Why were you initially drawn to philosophical issues concerning technology?

One of the most serious obstacles that philosophy of science and technology teachers have to overcome is the popular belief – sanctioned by most philosophers from Francis Bacon to Auguste Comte to Jürgen Habermas – that science and technology are the same thing: that there is no difference between physics and engineering, biology and medicine, psychology and psychiatry, economics and macroeconomic policy, and so on. This confusion has recently been given an academic name: "technoscience".

I was spared this confusion at an early age. My father, a physician among other things, was obsessed by tuberculosis – then a common affliction among the poor – and evolutionary biology. He had seen many TB victims as a medical student, and later on in his own bones. My father was also an enthusiastic Darwinist, as were all the educated and progressive men of his generation.

I still remember the dark and warm summer evening, possibly in 1924, when my father introduced me to evolutionary biology. He talked at length about our not distant cousins, the Madagaskar Lemurs. I understood intuitively that this subject was very distinct from the (then mostly unsuccessful) treatment of medical conditions. For one thing, whereas sick people were given pills and shots, fossil dinosaurs stayed quietly in books and, as I learned later on, in museums of natural history as well. Further, doctors and nurses hoped to be able to help the sick, whereas fossils were beyond help. Last, but not least, reconstructing the lives of fossils seemed to take just as much imagination as writing the novels of Malaysian pirates that the kids of my generation loved. In short,

I learned early on that one can study either for fun or to change the world around us. Today I put it this way: Whereas scientists study the world, technologists help alter it – for better or worse. Shorter: Science is about truth, technology is about utility.

Much later, when I entered the Universidad Nacional de La Plata to study physics, I shared all the subjects in the first two years with engineering students. I was part of a study group with four of them. Our differences in abilities and attitudes emerged soon after we first met. My partners were quick to imagine possible technological applications but hardly interested in the foundational problems that intrigued me. They had spatial imagination, which I lacked, and knew how to draw, whereas I was incapable of solving any problems in descriptive geometry. On the other hand I was adept at solving mathematical problems. In my third year our paths diverged: while my classmates studied machines, I studied atoms and things. Even our approaches to mechanics were different. Whereas they tackled difficult inverse problems of designing machines for performing desired functions, I faced the far easier direct problems of finding solutions of general equations in special cases.

I had a similar experience when teaching physics to electricians at the Workers' School that I founded and directed from 1939 on— and which the military government closed down at the end of 1943. I found that my worker-students were interested in understanding electric circuits, of which they had an intuitive comprehension, but found it hard to follow my popular presentation of the theory of alternating currents and Maxwell's electromagnetic theory, which I found movingly beautiful. Later on, when I taught courses on antennas and on wave guides at the Universidad de Buenos Aires, my physics and mathematics students were unimpressed: that stuff was far too near engineering to excite them. I suppose their reaction was partly genuine distaste for engineering, and partly snobbism and arrogance. I myself had the occasional bout of such snobbism and arrogance. For instance, when a philosopher friend and colleague enthused about science over the launching of the Sputnik in 1957, I said "Bah, that's only engineering!"

Another experience of the same kind was the philosophy course I taught at the engineering school at the request of the student union. (That course made one convert to philosophy: Alberto Coffa, who eventually studied at Pittsburgh and taught at U.C.) That forced me to pose and answer some of the most basic problems in the philosophy of technology, such as the nature of artifacts in contradistinction to natural things. As is well known, this question was first asked by Aristotle. And Marx gave the famous answer in his comparison of the bee with the architect: Whereas the bee builds the beehive instinctively, the house originates in the architect's brain. To generalize and indulge in a Platonic idiom: Unlike natural things, artifacts embody ideas.

The gist of my first serious foray into the philosophy of technology was my 1956 paper on Turing's problem, whether machines can think. This paper grew from a spirited discussion we had at my philosophical circle. My old friend, the mathematician Sadosky (who, like me, was then unemployed but three decades later was to become the first and last Argentinian minister of science and technology) was a recent enthusiastic convert to computer science – which was then called artificial intelligence. The idea that an artifact could think was attractive not only to engineers but also to philosophical materialists. This explains the continued popularity of the computer model of the mind: it looks like the least-expensive alternative to the idealist doctrine of the immateriality of the mind. My point was that computers do not think, but carry out physical operations that are made to correspond partially to mental operations, much as the balls in an abacus are feigned to be numbers. I illustrated this point by analyzing Pascal's ingenious mechanical computer.

Years later, when walking past a bookstore on my way to lecture on quantum mechanics at the Universidad Nacional de La Plata, I was attracted by Kimball's book on operations research. I entered the shop, leafed quickly through the book and bought it, spending a large fraction of my salary. I was fascinated by the prospect that OR would use the scientific method to pose and solve problems of large-scale human actions, such as shipping supplies across an ocean infested by enemy submarines, and keeping optimal inventories (neither too small nor too large) of manufactured goods. That book and some articles in a couple of journals on OR and management science reinforced my belief that the scientific method is topic-neutral: that it can be applied in all research fields. In other words, I became a convert to scientism, the thesis that the mathematician and politologist Marquis de Condorcet had stated brilliantly in 1769.

Beginning in the 1960s, scientism was attacked by the self-styled leftists in Western Europe and the Americas. The reasoning seems to have been this:

Technology is part of the Establishment.
The Establishment is bad.
We must fight the Establishment.

We must fight Technology.

Furthermore, since they believed that Technology = Science, the same people also resolved to fight science.

Ironically, the so-called New Left of the 1960s had adopted the same technephobia and epistemephobia that Heidegger had been preaching – making an exception for the German military technology, of course. The result was the same: to deprive individuals from the ability to think rationally, and governments from using the two most powerful levers to advance. In fact, this attitude provoked a flight of young Americans from science and technology, and the consequent deficit in these fields. Whereas in China, India, and Western Europe roughly one-third of undergraduates pursue careers in science or technology, in the US only about 17 percent do the same. Although I sympathized with the anti-war movement, I could not accept their wholesale rejection of science and technology, based as it was on the false identification of the two as well as on the failure to distinguish good from bad technology.

In 1961 I gave my first lecture on the philosophy of technology at the school of science of the Universidad de Buenos Aires. A revised and translated version of it appeared in the issue of *Technology and Culture* that has been credited with inaugurating the contemporary phase in the philosophy of technology. Unfortunately I allowed the editor, Melvin Kranzberg, to use the title of my paper for the whole issue, and to change my title into "Technology as applied science".

This was a mistake because the gist of any technological project is design. Surely, modern engineering, medicine, and the other technologies are scientific as opposed to empirical, but they are not sciences, whether basic or applied, because their ultimate goal is to design artifacts, not to find truths for their own sake. For example, physicists are occasionally curious about the way certain artifacts work, but few of them have designed any – except of course for experimental set-ups.

The only outstanding examples I remember off the cuff are Joseph Henry, who invented the electric motor; Ernest O. Lawrence, the designer of the cyclotron; and Enrico Fermi, who invented the nuclear reactor. They had an atypical mindset. In particular,

Fermi seems to have been the only physicist to do a calculation in the morning, work at the lathe in the afternoon, and sketch a part of an artifact in the evening.

2. What does your work reveal about technology that other academics, citizens, or engineers typically fail to appreciate?

Here is a sample of the ideas on technology that I have proposed and discussed in a number of publications, among them *Scientific Research* (1967, 1988), *Ciencia y desarrollo* (1980, 1997), *Treatise on Basic Philosophy*, Vol. 7, Part II (1985), and *Social Science Under Debate* (1996).

1/ Technology should not be confused with either craftsmanship or industry: Technology is a conceptual input to industry and government. In fact, every piece of technology – such as a blueprint or a description of an artifact or a man-made process – is a system of ideas, and consequently it raises plenty of philosophical problems glossed over by those who mix it up with either craftsmanship or industry.

2/ Whereas before the eighteenth century engineering owed little to science, from then on it depended critically upon it. For instance, electrical engineering uses Faraday's work on electromagnetism; telecommunications engineering utilizes the findings of electromagnetic waves; electronics is a practical offshoot of the discovery of the electron in 1898 and the subsequent theories of the electron computer technology would have been impossible without solid state physics, which in turn rests on quantum mechanics; the design of neural prostheses uses not only electronics but also neuroscience; and the most effective policies for controlling crime rest on criminal sociology.

3/ Contrary to popular belief, technology is much more than engineering: it includes all the disciplines employed in the rational alteration of reality. Thus, in addition to the various physical technologies, from civil engineering to computer engineering, there are industrial chemistry and chemical engineering, agricultural engineering and bioengineering, medicine and dentistry, education science, law, management science, normative macroeconomics, the policy sciences, and even practical philosophy, in particular ethics and political philosophy.

4/ Technology does not flow spontaneously from basic science, and it is not the same as applied science either. For example, bridge design makes use of theoretical mechanics (a basic science), and materials science (an applied science), but it is contained in neither of them. It is up to the engineer to tackle the problem of bridging the chasm in the most efficient way – rather than, say, by sheer bulk and manual labor, the way the Roman engineers did.

5/ There is a radical difference between original design and improvement. An improvement follows the use of a previous invention. By contrast, a radical innovation – such as that of the electric motor, the electric bulb, penicillin, the genetically modified organism, the automatic teller, or microfinancing – has no precedents. Consequently, Henry Petrosky, in his otherwise admirable books on the history of technology, is wrong in claiming that all technological innovation comes from correcting mistakes in previous artifact design. A mistaken flash of genius must occur before it can be criticized. The engines of technological (and scientific) creativity are curiosity and disciplined imagination: criticism is only a quality-control mechanism.

6/ Every technology raises philosophical problems of several kinds. For example, seismic engineering poses the epistemological problem whether it is valid to use subjective probabilities (of, e.g., the occurrence of an earthquake), and the moral problem of optimizing safety while keeping costs down. And the artificial fertilization techniques pose the problem whether it is fair to the child to be raised by a sexagenarian mother. Technoethics is well on its way, albeit sometimes disturbed by obsolete ideological concerns justifiable at times and places where the priestly caste was powerful.

7/ If technology is the discipline concerned with the design and maintenance of artifacts and processes of some kind, then it must be admitted that philosophy has always contained a technology of its own, namely practical philosophy, along with theoretical philosophy (such as logic and ontology) and the history of the discipline. I submit that practical philosophy today consist of axiology (value theory), praxiology (action theory), ethics (moral philosophy), and political philosophy. All of these are technologies because they deal, just like human engineering and education science, with the control of human behavior. Being technologies, they should be judged not only as to their foundations if any but also as to their practical fruits.

8/ Basic science is morally and ideologically neutral: Scientific findings are true or false, not good or bad. By contrast, technology is morally compromised, because its ultimate products, namely artifacts, can be put to either prosocial or antisocial uses. For example, the design of WMDs and sterile seeds are morally objectionable, that of vaccines and dynamos moved by small waterfalls is not. Still other artifacts, from knives to nuclear plants, are morally ambivalent. As for the technology-ideology connection, it is just silly to repeat, after Habermas, that science and technology (which he identifies) constitute the ideology of late capitalism. Evidence, please, Herr Professor, evidence!

9/ Basic scientific research should be free from political interference, which can only throttle it. But, because there are intrinsically evil artifacts, such as offensive weapons, and ambivalent ones, such as Internet (which can be used not only to facilitate constructive communication but also to organize criminal gangs), democratic governments and NGOs have the duty to watch technological developments and hold hearings about their prospective criminal uses. Shorter: Frisk the technologist, not the basic scientist.

10/ Every policy and every plan of national development should include strong support for science and technology, because one of the traits of underdevelopment is precisely a severe deficit in both. In the case of technology, preference should be given to local needs. By contrast, in the case of science local talent should be prioritized. A poor country cannot afford nuclear engineering or astronautics, but it can certainly afford agronomists, engineers, and management scientists, as well as mathematicians, historians and anthropologists – depending on the local talent pool and educational facilities.

3. What, if any, practical and/or social-political obligations follow from studying technology from a philosophical perspective?

Let me to begin by drawing a distinction between genuine philosophy, such as Aristotle's, Spinoza's, or Kant's; protophilosophy, such as the pre-Socratics's, Ludwig Wittgenstein's, Donald Davidson's or John Searle's; and pseudophilosophy, such as Heidegger's, Gadamer's, or Derrida's. Unlike protophilosophy, genuine philosophy tackles important problems with the help of advanced conceptual tools and in the light of scientific knowledge. And, unlike pseudophilosophy, authentic philosophy is clear and subject to rational debate. It aims at discovering truths, not at discrediting the very notion of truth or advancing special interests. Moreover, genuine philosophy is universal, not tribal: it can be done anywhere for the delight of anyone. And, like technological research, philosophical research can be more or less rigorous and original. Thus, just as there are maintenance engineers, medical practitioners, and barristers along with design engineers, biomedical researchers and jurists, so there are also uncounted philosophy teachers for every original philosopher.

The study of technology from a philosophical perspective should lead to a reconceptualization of modern culture, and therefore to a reformulation of cultural policies. Indeed, when one realizes that every technology is a body of ideas, one understands that it belongs in culture along with mathematics, basic science, the humanities, and art. In other words, technologists are intellectuals rather than craftsmen. Moreover, what they do can change industry or government, hence society, in a more lasting fashion than military feats. Hence any definition of 'modern culture' that fails to include technology is hopelessly dated and bound to inspire ineffective or even counterproductive cultural policies.

I submit that a progressive cultural policy will include strong support for technological development and education. And, since technology – whether appropriate or advanced – depends critically upon scientific discovery, such a policy will also include strong support for basic science and applied science – the latter being the bridge between basic science and technology. Without these two, no developing nation can hope to go beyond the stage of mere supplier of natural resources and unskilled workforce to the highly industrialized world.

And yet, because most of the development programs in underdeveloped countries are designed by economists trained at Harvard, Chicago, or their local imitators, they continue to regard science as a luxury that poor countries cannot afford. Those economists do not wish to learn what neither Adam Smith nor David Ricardo, nor even Karl Marx or Alfred Marshall, could have anticipated, because they were blinded by their Eurocentric perspective: that poor countries must develop their own industries instead of continuing to exchange low-priced foodstuff like soja beans, and commodities like lumber, for high-priced manufactured goods such as cars. Nor do those teleplanners know that genuine and lasting development, far from coinciding with exports growth, involves cultural (in particular scientific and technological) progress along with genuine democratization.

4. If the history of ideas were to be narrated in such a way as to emphasize technological issues, how would that narrative differ from traditional accounts?

A first difference would be that, instead of starting in the sixth century B.C., it would start in the eighteenth A.D., which is when the first technologists proper lived: Leonhard Euler, the founder of naval engineering; James Watt, the inventor of the first negative feedback control device; and Lazare Carnot, the first to study the thermodynamics of steam engines. Before them there were plenty of ingenious craftsmen, but no technologists in the modern sense of the word: they worked by trial and error instead of using mathematics, theoretical mechanics or other scientific theories to solve practical problems. Thus the designers of the first steam engines knew the physical principles of mechanical engineering (the mechanics of Newton and Euler), but they knew very little about heat, and that little was only half-true, since it included the false hypothesis that heat was a fluid.

However, good historians of ideas are supposed to narrate not only what people thought before them, but are also expected to explain why certain opportunities were missed. (For instance, why did the Scientific Revolution not happen in India, the cradle of so many subtle thinkers? And why did China, with so many ingenious inventors and bold explorers, fail to become a naval power?) I suppose that the alternative history in question would also explain why until recently philosophers have been – as Marx pointed out – more interested in "interpreting" the world than in changing it.

5. With respect to present and future inquiry, how can the most important philosophical problems concerning technology be identified and explored?

Here is a quick and haphazard list of problems that, in my opinion, deserve being explored in greater depth and detail.

1/ The vast majority of design problems are inverse (or synthesis) problems rather than direct (or analysis) problems. In fact, they are of this type: Design a gadget that will perform a given function. Equivalently: Find the mechanism and the input to it that will result in a desired output. The corresponding direct problem is of the form: Given a mechanism and the input to it, calculate the output. Unsurprisingly, the general study of inverse problems has been mushrooming over the past decade in physics and engineering, to the point that there are several new journals and handbooks devoted to them. But, once again, philosophers have lagged behind. A few years ago a paper of mine on inverse problems was rejected by several philosophy journals because none of the referees had ever heard the very expression "inverse problem", which is of course familiar to mathematicians, quantum physicists and engineers. Eventually the paper became two chapters of my *Chasing Reality* (2006).

2/ Would it be possible to construct computers capable of creating new concepts, posing new problems, devising new algorithms, and making value judgments in new circumstances? The problem is not to reheat the many extravagant promises made by computer fans since 1950; nor is it to reject these promissory notes out of hand. The problem is to give plausible reasons why only brains in intellectually stimulating and tolerant environments can possibly come up with original ideas, problems, and value judgments, such as "Democracy is better for the demos than aristocracy".

3/ Perform detailed studies of normative epidemiology, the law, management science, social work, and education science as sociotechnologies (or branches of social engineering).

4/ Should the governance of social systems include technological inputs? If so, would technology circumvent the need for rational debate and democratic decision-making? Assuming that technological development should be placed under democratic control, how could such control be implemented without suffocating potentially useful innovations?

5/ Does technocracy have irremediable flaws, and if so which are they: political, moral, or both?

As for how any of these problems could best be approached, respondeo: Against a broad and clear philosophical background (including lots of good metaphysics), and with the help of some relevant scientific and technological knowledge. Take, for instance, the last of the problems listed above, the one about technocracy. I suggest that one should start by answering the radical anarchist claim, that people can organize themselves without any control or coercion. A study of complex living and social systems, such as cells and factories, should show that their regular functioning requires controls to face unexpected external disturbances (such as tsunamis and currency devaluations) and disintegrating internal forces (such as egoism and complacency).

But one should also own that, since every technological project aims at altering people's lives, and since some such changes may be harmful, technologists should not be trusted to rule without consulting the views and interests of the people most likely to be affected. In other words, in this case, like in any case involving complex systems, one should adopt a systemic approach: one should place the artifact in its social setting. In short, technocracy cannot work to everyone's benefit because the technologist tends naturally to focus on the artifact, sometimes with disregard for its users

However, I submit that a modicum of technological knowledge and concern for technoethics are not sufficient to tackle correctly any problems in the philosophy of technology. One should also make use of the basic philosophical disciplines in their present state – logic, semantics, ontology, epistemology, and ethics.

Consider, for example, this problem in political philosophy, which I regard as part of philosophical technology: Which is the most fair (moral) of all property regimes: monopoly, free market, statism, or cooperativism? I suggest that this problem is parallel to the one concerning government: Which governance procedure is the best: autocracy, oligarchy, tyranny, or democracy? In both cases the problem is not only one of practical efficiency but also moral, because it boils down to this: Under which regime are the many more likely to enjoy life and at the same time be motivated to help others live? Since the reader must have guessed what my answer is, I feel justified in closing by stating a profound platitude: Technology without morals can be evil, and ethics without technology is impotent.

Selected Bibliography

1. *Causality: The Place of the Causal Principle in Modern Science.* Cambridge, Mass.: Harvard University Press, 1959. Last ed.: Dover, 1979. Seven translations.

2. *Metascientific Queries.* Springfield, ILL.: Charles C. Thomas, Publisher, 1959.

3. *Intuition and Science.* Englewood Cliffs, N.J.: Prentice-Hall, 1962. Two translations.

4. *The Myth of Simplicity.* Englewood Cliffs, N.J.: Prentice-Hall, 1963.

5. *Scientific Research I: The Search for System.* Berlin-Heidelberg-New York: Springer-Verlag, 1967. Rev. ed. *Philosophy of Science* 1. Transaction, 1998. One translation.

6. *Scientific Research II: The Search for Truth.* Berlin-Heidelberg-New York: Springer-Verlag, 1967. Rev. ed. *Philosophy of Science* 2. Transaction, 1998.

7. *Foundations of Physics.* Berlin-Heidelberg-New York: Springer-Verlag, 1967.

8. *Philosophy of Physics.* Dordrecht: Reidel, 1973. Five translations.

9. *Method, Model and Matter.* Dordrecht: Reidel, 1973.

10. *Sense and Reference. 1st vol. of Treatise on Basic Philosophy.* Dordrecht: Reidel, 1974. One translation.

11. *Interpretation and Truth. 2nd vol. of Treatise on Basic Philosophy.* Dordrecht: Reidel, 1974. One translation.

12. *Tecnología y filosofía.* Monterrey, México: Universidad Autonoma de Nuevo Leon, 1976

13. *The Furniture of the World. 3rd vol. of Treatise on Basic Philosophy.* Dordrecht: Reidel, 1974.

14. *A World of Systems. 4th vol. of Treatise on Basic Philosophy.* Dordrecht: Reidel, 1979.

15. *The Mind-Body Problem.* Oxford and New York: Pergamon Press, 1980. Two translations.

16. *Ciencia y desarrollo.* Buenos Aires: Siglo Veinte, 1980. One translation.

17. *Scientific Materialism.* Dordrecht-Boston: D. Reidel Publ. Co. 1981. Two translations.

18. *Exploring the World. 5th vol. of Treatise on Basic Philosophy.* Dordrecht: Reidel, 1983.

19. *Understanding the World. 6th vol. of Treatise on Basic Philosophy.* Dordrecht: Reidel, 1983.

20. *Philosophy of Science and Technology, part I: Formal and Physical Sciences. Part of Treatise on Basic Philosophy ,Vol. 7* Dordrecht: Reidel, 1985.

21. *Philosophy of Science and Technology, part II: Life Science, Social Science and Technology Part of Treatise on Basic Philosophy. , Vol. 7.* Dordrecht: Reidel, 1985.

22. *Philosophy of Psychology* (with Rubén Ardila). New York: Springer-Verlag, 1987. Two translations.

23. *Ethics: The Good and the Right, Vol. 8 of Treatise on Basic Philosophy.* Dordrecht-Boston: Reidel, 1989.

24. *Finding Philosophy in Social Science.* New Haven, CT: Yale University Press, 1996. One translation.

25. *Foundations of Biophilosophy* (with Martin Mahner). Berlin-Heidelberg- New York: Springer-Verlag, 1997. Three translations.

26. *Social Science Under Debate.* Toronto: University of Toronto Press, 1998. One translation.

27. *Dictionary of Philosophy.* Amherst, NY: Prometheus Books, 1999. Rev. ed." Philosophical Dictionary, Prometheus, 2001. Two translations.

28. *The Sociology -Philosophy Connection.* Foreword by Raymond Boudon. Brunswick, NJ: Transaction Publishers, 1999. One translation.

29. *Philosophy in Crisis: The Need for Reconstruction.* Amherst, NY: Prometheus Books, 2001. One translation.

30. *Scientific Realism: Selected Essays by Mario Bunge.* Ed. Martin Mahner. Amherst, NY: Prometheus Books, 2001.

31. *Emergence and Convergence.* Toronto: University of Toronto Press, 2003. One translation.

32. *Chasing Reality.* Toronto: University of Toronto Press, 2006. One translation.

4

Harry Collins

Distinguished Research Professor

Cardiff University
UK

---

1. Why were you initially drawn to philosophical issues concerning technology?

I wasn't: I'm a sociologist. As it happens, however, for me, the disciplinary boundary between sociology and philosophy is not as great as it might be. One reason is that those who, like me, studied sociology in Britain in the 1960 and 1970s, found themselves dealing with quite a lot of philosophy. Philosophy of the social sciences, and of the sciences, had a big role in the sociology syllabuses of 1960s Britain.

The boundary is still less marked for a sociologist of science, particularly a sociologist of scientific knowledge since a lot of the time we are dealing with questions about the nature of knowledge. The sociology of scientific knowledge (SSK as it is widely known), tries to extend the sociology of knowledge, pioneered by Mannheim, to scientific knowledge. We therefore ask questions such as why do some scientists believe that, say, gravitational waves have been detected whereas others do not, just as a regular sociologist of knowledge might have asked why some people believe in witches and some people do not believe in witches. The ordinary sociologist of knowledge tends to locate the causes of differences in individual belief in the society and, likewise, the sociologist of scientific knowledge tends to think of scientific knowledge as belonging to societies rather than individuals; the influences include the sociologist, Durkheim who said 'treat social facts as things,' and the philosopher Wittgenstein with his idea of 'form-of-life.' Indeed, sociologists of scientific knowledge tend to think of Wittgenstein as a kind of sociologist, making the sociology-philosophy boundary still more insecure. It has turned out, furthermore, that I have co-authored a book with a philosopher and that I have published in philosophy journals and occasionally get asked to present my work at philosophy conferences. It has to be admitted, however, that my aim can be seen as subversive as I want to replace the traditional philosophical answers to epistemological questions with sociological answers. In particular, I start with the idea that the prime location of knowledge is the collectivity rather than the individual.

2. What does your work reveal about technology that other academics, citizens, or engineers typically fail to appreciate?

I think that the answer more regular sociologists of technology would give to this question would differ from mine. They would say that their studies show the importance of social understandings of technology to the technological trajectories of artifacts; they would say that technology is 'socially constructed.' To repeat a standard example due to Pinch and Bijker (Bijker, Hughes and Pinch, 1987), they would say that the technical evolution of the bicycle mostly had to do with whether the bicycle was seen as a dangerous and sexy machine which enabled young men to impress their girlfriends or as a safe and efficient means of transport for everyone. This was more important than the 'logic' of the machinery. My work on science, notably what has become known as the Empirical Programme of Relativism (EPOR – Collins 1985/92) has been used as an input to this kind of thinking.

Where there is an overlap with the regular sociologists of technology, notably Trevor Pinch, is in the application of another part of sociology of scientific knowledge to technology. Actually, I have always claimed that there is little distinction between the sociology of science and the sociology of technology because all the science I know looks pretty technological. For example, for more than thirty years I have been studying physicists' attempts to detect gravitational waves. What these physicists do in their day-to-day lives is design, build, and 'debug' ever bigger and more complex machines. Thus in my big sociological history of the enterprise (Collins, 2004a) I say that among the problems I deal with in the book are:

> ... the triumph of one technology over another; the necessity for choice without complete justification; the pruning of the potentially ever-ramifying branches of scientific and technological possibility if science is ever to move forward; ... and the tension

between the world seen as an exact, calculable, plannable sort of place, just waiting for us to get the sums right, and the world seen as dark and amorphous, bits of which, from time to time, we are lucky enough to catch in our speculatively thrown nets of understanding—if we throw them with sufficient skill. (p18)

The last question has led Pinch and I to collaborate on a book about technology (Collins and Pinch 1998), in which a number of the case studies turn on the uncertainty of science and technology as it is lived out. One case study looks at attempts by agencies to prove the safety of technologies by staging spectacular public demonstrations. Examined closely, it turns out if a demonstration looks too convincing it is precisely because of the way it has been 'staged;' in real life matters are far less clear so that neither the test of the safety of an anti-misting kerosene for aircraft or of a nuclear fuel flask which was subject to a train crash, both of which we describe, proved what they appeared to prove. Another study showed how it was that during the Gulf War it was impossible to tell whether Patriot missiles had shot down all or none of the Scud rockets fired by Iraq. A third study looks at the crash of the Challenger space-shuttle. Here we find ourselves at odds with philosophers of technology such as Andrew Feenberg. Feenberg argues, along with many others, that the tragedy was the result of managers pressurizing engineers to stay on schedule even though the engineers realized the launch was dangerous. Pinch and I argue that this is to misconceive the problem. If you understand the 'amorphousness' and 'darkness' of the world of frontier technology you cannot escape the conviction that no space shuttle could ever be safe enough to merit the symbolic accolade of carrying a school teacher into the heavens. Astronauts know they are taking risks with their lives whereas flying a school teacher says 'this machine is as safe as an airplane.' And yet the two crashes we have seen in 100 or so shuttle flights is close to what the risk assessors calculated at the outset of the programme. To put it another way, sitting on top of hundreds of tons of explosive and setting light to it is never likely to be as safe as flying the Atlantic in a jet, especially as aircraft have been developed for a hundred years and jets are nowadays tested almost to destruction before they are allowed to carry passengers. The sociological puzzle is not how the managers managed to over-rule the engineers – our argument in the paper is that this is likely to happen frequently if the engineers are doing their job properly and yet flights are still going to be allowed – but how NASA came to fly the school teacher in the first place? The interesting philosophical question here, is how is it that so many of our colleagues are ready to play the 'blame game,' their premise being that the machine was potentially safe enough to do the job the politicians asked it to do? Why are analysts of technology buying President Reagan's story?

These studies aside, when it comes to technology narrowly conceived, I have been using my findings on the nature of scientific knowledge to analyse artificial intelligence. Here my input has not been that of a social constructivist but rather of a contributor to the field – showing that current ambitions are bound to fail. I have argued that since most knowledge is essentially social – it is society that is responsible for developing, affirming and maintaining knowledge – until machines can become members of society they won't achieve anything that looks like human knowledge (see, for example, Collins 1990).

The point is quite easy to demonstrate because the problem of AI has a 'holographic' quality. By this I mean that the entire problem can be illustrated by looking at only a small 'splinter' broken out of the whole pane. For instance, we can look at spell-checking.

Consider the following sentence: 'My spell-checker will flag weerd processor but won't flag weird processor.' Now, what is claimed in that sentence is true. As I typed that sentence on my word processor, using Word for Windows, the word 'weerd' was underlined with a spiky red line whereas the word 'weird' was not. And, in fact, it keeps happening as I type 'weerd' repeatedly. The problem here is not to get the spell-checker to flag 'weird processor.' That is something we could easily achieve by having the dictionary deal with two-word sequences as well as single words. The problem is to get the spell-checker not to flag 'weerd.' I don't want it to flag 'weerd' because 'weerd' is spelt just as I want it to be spelled throughout this passage. To know that 'weerd' should not be flagged requires all the intelligence it takes to understand the passage. 'Weerd' has been flagged eight times in this passage whereas a spell-checker that functioned as well as a human sub-editor would not have flagged it at all. To reach that level of understanding the spell-checker is going to have to learn spelling by being embedded in society in the same way as the sub-editor: social embedding in the language seems to be how we learn to understand complex passages like this one.

The other thing that I consider to be a really deep insight, but one that hardly anyone has picked up, is that though nearly all knowledge is social there is a class of knowledge that, effectively, is not (Collins and Kusch, 1998). To see this one has to talk of actions rather than knowledge. The starting point is that humans are so competent that if they choose they can act like non-human entities. Thus, I can stand completely still on a stage with my arms held out and imitate a tree. In so far as I can imitate a tree, a tree can imitate me. To show that this is not just fantasy we might note that these days, in our city centres we see people earning money by standing completely still pretending to be statues. A visitor from another planet might not be sure whether it was the people in the city centre who were imitating the statues or the statues who were imitating the people. This is imitation of non-human entities, which Kusch (who is a philosopher), and I call mimeomorphic action. It is action that can mimicked by reproducing the behaviour associated with action without taking account of the intentions behind it. Therefore, machines without intentions or any other kind of social understanding can mimic mimeomorphic actions merely by imitating behaviour. This cannot be done in the case of the other kind of actions – polimorphic actions – because to repeat the same polimorphic action over and over again normally requires different behaviours in different social contexts (think of the action of greeting). My claim is that we can only understand the way machines fit into society – the way they function as 'social prostheses' (Collins, 1990) – if we break up our own actions into their polimorphic and mimeorphic components first. There is much more to it than this, but that is a basic requirement.

3. What, if any, practical and/or social-political obligations follow from studying technology from a philosophical perspective?

In this regard my work tends to go against the modern trend of much of social studies of science which stresses the importance of the lay public in technological decision making. Consider the cases mentioned above. Analysis of the attempt by British Nuclear Fuels to demonstrate the safety of their fuel flasks, and the attempt by the FAA to demonstrate the safety of anti-misting kerosene, shows that to understand what underlay the convincing-looking appearances you had to know quite a bit more about the tests than the viewing public knew. In fact, examined carefully, neither test showed what it appeared to show.

The correct analysis of the Challenger tragedy, it seems to me, also requires one to understand quite a bit about technology; it requires one to have some sense about how safe and unsafe various technologies really are. In that case even some very clever social analysts have followed the politicians and fallen into the trap of treating the space shuttle like an airplane in terms of safety. To know it you need to know something of how different kinds of machine work and how they are safety-tested – this is not knowledge that is universally available.

Of more widespread importance is something like the argument about the safety of vaccination. In Britain in the early 2000s a panic set in among the population about a triple vaccine for Mumps, Measles and Rubella – the MMR vaccine. As a result of the irresponsible claims of one doctor, made during a press conference, parents stopped taking up the MMR vaccine. The result was the beginnings of an epidemic of measles – a dangerous disease. There is little doubt that lives were lost and permanent injury caused because of the panic. A quick but careful examination of events showed that there was simply no scientific evidence to back up the irresponsible doctor's claim yet, once more, journalists, the public, and a disappointingly large proportion of the social science community, took the side of the protesting parents almost as a reflex reaction. Within much of that part of the social science and humanities community who take their role as commentary upon science, there seems to be a romantic attachment to the views of the general public and a prejudice against the claims of scientists and technologists – in our latest book (Collins and Evans 2007), we call this the 'folk wisdom view.'

As we argued a few years ago (Collins and Evans 2002), the initial weakening and leveling down of the claims from authority typically made by scientists and technologists in the 1950s was a healthy development. The social sciences and the humanities took a leading role in this leveling down. Furthermore, given the visibility of so many technological tragedies it was inevitable that the top down authority of 1950s science had to be tempered by wider consultation with the public. Without such widespread consultation, radical new technological developments affecting the public would cease to have political credibility. In our 2002 paper we referred to the growing consultation with the public a solution to the 'Problem of Legitimacy.' We argued, however, that in so far as more and wider consultation solved the Problem of Legitimacy it left in its wake the 'Problem of Extension.' No one knew how to limit the role of the public in technological decisions-making even where the public did not understand the problems. Constructivist and postmodernist critiques of science and technology left us without a way of saying that anyone 'did not understand a problem' since there was no longer anything fixed to understand – all was politics in the disguise of science. We argued that a solution to the Problem of Extension was vital and that this solution had to come through a better analysis of expertise and experience. We argued that scientific truth as conventionally understood was as brittle as glass and its claims would always be found wanting when confronted with technological reality. What was needed was a model of scientific knowledge that was mixed, ugly, but resilient – like reinforced concrete. We argued that expertise and experience could fill this role and we suggested a new programme of research: Studies of Expertise and Experience (SEE). Since then we have begun that programme and the first results are reported in our 2007 book and continually updated on our website, www.cf.ac.uk/socsi/expertise

One of the ambitions of SEE is to say something about who is entitled to contribute to the technical component of a technological decision. We have argued in the works referred to that to make a sensible contribution requires some understanding of the technology, but how much understanding? We can say for sure that scientists and technologists who work in the areas in question have enough understanding to contribute but to restrict contributions to these alone is to go back to the old top-down model associated with Polanyi's 'Republic of Science' and this no longer seems tenable, not to mention its legacy – the Problem of Legitimacy. On the other hand, we don't want to go all the way in the other direction and end up with the Problem of Extension and the 'folk wisdom view' (the complement of the Republic of Science).

SEE tries to work out what kinds of in-between expertises there are and to start some discussion of which of these might or might not be reasonably taken to put their possessors in a position to contribute sensibly to technological debates. We have tried to begin this discussion by setting out what we call the Periodic Table of Expertises. We are not sure that the Periodic Table is the finishing point in the classification of expertise but it is a starting point. Perhaps it will last but perhaps its value is as a provocation – something that will get philosophers and others to think about the problem. Irrespective of which way it turns out, we have begun to develop philosophical (Collins 2004b) and empirical (Collins et al 2006) research based on the Table. Others are joining us in this critical and empirical work (Collinsd 2008). We have done most of our work so far on one of the distinctions – that between interactional and contributory expertise – but the analysis of more areas is being developed.

Interactional and contributory expertise

Contributory expertise is easy to explain because it is simply what most people mean by expertise and is certainly what Polanyi had in mind. It is the expertise required to contribute, practically, to the domain of expertise in question. Thus, a contributory expert in the domain of gravitational wave detection physics is someone who can make a contribution to that field such as by designing suspensions for interferometer mirrors, calculating the gravitational wave waveforms produced by various sources such as inspiraling binary neutron stars, or analyzing the data produced by the detectors to determine whether a signal can be extracted from the noise in any particular case. The idea of interactional expertise, on the other hand, is most readily exemplified by the skill required by a sociologist of scientific knowledge, working in the domain of gravitational wave detection, who learns to engage respondents in reasonably high level of technical discussions of the scientific domain yet still without being able to publish or carry out experiments in the domain; the sociologist develops interactional expertise without developing 'contributory expertise.'[1]

Interactional expertise should not be confused with the kind of formal knowledge that can be written down. Interactional expertise involves mastery of a language and is characterized by rules that cannot be explicated; it is a tacit knowledge-laden ability that can only be acquired through immersion in the discourse of the community. It is usually thought that the tacit components of the language of a specialist domain can be acquired only by full immersion in the esoteric form-of-life including its practical components. We, however, have done experiments which suggest

[1]For a philosophical discussion of interactional expertise which relates it to existing philosophical ideas and is illustrated with examples from fiction and psychology see Collins 2004b; for a discussion of interactional versus contributory expertise in the context of sociological research on scientists, see Collins 2004a; for types of expertise in the context of relations between scientists and the public, see Epstein, 1996, Collins and Evans 2002, and Collins and Pinch 2005. For a more complete classification of expertises see 'The Periodic Table of Expertises' and the exposition in Collins and Evans 2007 and at www.cf.ac.uk/socsi/expertise.

otherwise. We define the 'strong interactional hypothesis.' This states that a person with maximal interactional expertise will be indistinguishable from a person with interactional and contributory expertise in any test conducted by verbal means alone. In our experiments we tested the conversational abilities of colour-blind persons – who we took to have maximal interactional expertise in colour-perception since they had been embedded in the discourse all their lives – against the conversational abilities of persons who could see colours. We found that colour-perceiving judges could not distinguish between them, supporting the strong interactional hypothesis. We then went on to show that gravitational wave physicists could not distinguish Collins (who has been immersed in the gravitational wave field for about 30 years but is not a physicist), from other gravitational wave physicists (Collins et al 2006). If these experiments show what they appear to show, then immersion in the sea of language alone, in the absence of practical experience, is enough to enable the entire language to be acquired.

Interactional expertise does not provide the ability to do anything practical within the domain, nor, arguably (see work on the function of mirror neurons), does the interactional expert share the experiences of what it is to be practically involved in the domain, but this does not adversely affect the interactional expert's ability to use the language fluently.[2] The language is the same as the language of one who has shared the experiences. It follows that the interactional expert can make sound judgments and decisions that bear on practical matters.

Using Merleau-Ponty's well-known example so as to express the argument as starkly as possible, the position developed here would imply that seeing-people without sticks can, in principle, fully acquire the language of blind people with sticks if they talk to them long enough. Furthermore, such seeing-people without sticks would be just at good as blind people with sticks at advising stick-manufacturers on the best kind of sticks to make for blind stick-using people and at guessing what it would be like to use some new kind of stick – such as one that vibrates – even though no-one had ever held such a stick.[3] Of course, such a seeing

---

[2] I say arguably because it has been suggested by Theresa Schilhab (in preparation) that the functioning of mirror-neurons could provide some of the experience too.

[3] The challenge of a vibrating stick was suggested by Bert Dreyfus in private conversation.

person would not be able to try the experiment with authenticity and report what it really feels like to hold a vibrating stick, but once vibrating sticks had become common in the blind community the seeing person would be able to master the vibrating-stick discourse and discuss their relative merits as effectively as blind people. Interactional expertise stands between the idea of knowledge supported by, say, proponents of artificial intelligence, who think that all knowledge is potentially encodable, and that of phenomenological philosophers, who think that practice is the essence of all understanding.

The crucial point is that the uses of interactional expertise go well beyond research in the sociology of scientific knowledge. Restricting our gaze to science and technology, interactional expertise is often the medium of specialist peer review in funding agencies and in journal editing, where the reviewers are only sometimes contributors to the narrow specialty being evaluated. It is the medium of interchange within large scale science projects, where again not everyone can be a contributor to everyone else's narrow practical specialty. It is, *a fortiori*, the medium of interchange in properly interdisciplinary, as opposed to multidisciplinary, research. And, most important, on those occasions when activists or other concerned persons are driven to it, it can be the medium of interchange between scientists and groups of citizens (e.g Epstein 1996, Collins and Pinch 2005). Thus, the acquisition of interactional expertise puts citizens in a position to contribute to scientific judgments.

This is just one example of how the research programme of SEE is meant to bridge the gap between the Republic of Science and the folk wisdom view. What is needed next is work out the contribution that can be made by other types of expertise found in the Periodic Table.[4]

4. If the history of ideas were to be narrated in such a way as to emphasize technological issues, how would that narrative differ from traditional accounts?

I believe there would be little difference given that science and technology have so much overlap. The crucial difference to the

[4]We begin to ask about the expertise of the managers of big science projects in Collins and Evans 2007 and develop it in Collins and Sanders, 2008. Collins, 2008, contains papers that develop other elements of the periodic table too.

history of ideas that would emerge from work like mine would be to stress the collective locus of all knowledge.

5. With respect to present and future inquiry, how can the most important philosophical problems concerning technology be identified and explored?

As Humphrey Littleton is supposed to have said, 'If I knew where jazz was going I'd be there already.' I do not think you can sit down to identify problems, they just have to occur to you. One can, however, turn the question on its head and ask how one could make sure that the important questions are not identified and explored. If you can answer that it might give some guidance in respect of the more imponderable question.

The answer to the question of how to make sure that important questions are not identified is to treat philosophy of technology as a subject that is about the contents of philosophy of technology books rather than about the technological world. The same thing is true of philosophy in general. Compare the philosophy of someone like Russell who admits to burning out his intellect trying to complete the translation of mathematics into logic, or Wittgenstein, who tortured himself to the point of madness in trying to work out the relationship between words and things, with the phalanxes of 'professional philosophers' who are brilliant scholars but do not seem to be terribly interested in how the world works. Of course, our education system encourages this kind of thing because university examinations in the humanities are nearly always tests of scholarship and criticism, not of creative observation. So, working backwards, to identify the important problems in philosophy of technology it is probably a good idea to get out of the library and spend a lot of time hanging around technology. That at least opens up the possibility that interesting new questions will arise.

I would say that 95% of the good work I have done in the sociology of science has arisen quite unexpectedly as a result of hanging around science and scientists. All the really interesting things have turned out to have little to do with any hypotheses I took into the field and some of them arose out of stumbling over my own mistakes. Someone with a bit more foresight and ability to plan their work sensibly would not have discovered them and someone who stayed in the library would have had no chance. Mistakes, or unforeseen problems, are to be treasured because you can't have mistakes or problems unless you are engaging with something outside your own prejudices. New ideas often come from finding ways of overcoming mistakes or finding solutions to what seem to be intractable problems. So you have to maximize your chance of trouble by putting yourself in the way of it. It's a bit like being run over – stay on the sidewalk and you'll be safe but you won't get that deeper vision that might come from spending some time in a changed state of consciousness.

That said, let me break my rule and say that I think today's problems are all about the balance between expertise and social responsibility – if I have to name a place to look, that's where it would be.

Bibliography

Bijker, W., T. Hughes, T., and T. Pinch Eds (1987). *The Social Construction of Technological Systems.* Cambridge, Mass.: MIT Press.

Collins, H. M. 1985. *Changing Order: Replication and Induction in Scientific Practice.* Beverley Hills & London: Sage [2nd edition 1992, Chicago: University of Chicago Press].

Collins, H. M. 1990. *Artificial Experts: Social Knowledge and Intelligent Machines.* Cambridge, Mass: MIT press.

Collins, Harry 2004a. *Gravity's Shadow: The Search for Gravitational Waves.* Chicago: University of Chicago Press.

Collins, H. M. 2004b. "Interactional Expertise as a Third Kind of Knowledge." *Phenomenology and the Cognitive Sciences 3 (2)*:125-143.

Collins, H. M., and Robert Evans 2002. "The Third Wave of Science Studies: Studies of Expertise and Experience." *Social Studies of Science 32 (2)*:235-296.

Collins, Harry, and Robert Evans (2007 forthcoming). *Rethinking Expertise.* Chicago: University of Chicago Press.

Collins, H. M., R. Evans, R. Ribeiro, and M. Hall 2006. "Experiments with Interactional Expertise." *Studies in History and Philosophy of Science, 37A (4)*:000-000 [December].

Collins, H. M., and M. Kusch 1998. *The Shape of Actions: What Humans and Machines Can Do.* Cambridge, Mass: MIT Press.

Collins, H. M., and T. J. Pinch 1998. *The Golem at Large: What You Should Know About Technology.* Cambridge & New York: Cambridge University Press.

Collins, Harry, and Trevor Pinch 2005. *Dr Golem: How to think about medicine.* Chicago: University of Chicago Press.

Collins, Harry and Gary Sanders, 2008. "'They give you the keys and say drive it!' Managers, referred expertise, and other expertises," in Collins 2008.

Collins, Harry, ed., 2008 (forthcoming), *Studies of Expertise and Experience*: *Special Issue of Studies in History and Philosophy of Science.*

Epstein, S. 1996. *Impure Science: AIDS, Activism and the Politics of Knowledge.* Berkeley, Los Angeles & London: University of California Press.

5

Paul Durbin

Department of Philosophy

University of Delaware
USA

---

1. Why were you initially drawn to philosophical issues concerning technology?

For me the move from a focus on philosophy of science to philosophy of technology was a natural development – from science to applied science to technology (in the now-outdated terminology of the time). I was finishing up my doctoral thesis on the discovery process in science – later published as *Logic and Scientific Inquiry* (1968). My thesis (recall that a doctoral thesis is exactly that) was that plausible reasoning is the key to understanding the discovery process in science. No one today doubts the fundamental importance of probability and statistics in the whole range of contemporary sciences, but every application I know of in real-world science involves non-certain, plausible reasoning. While writing my thesis, I grew increasingly interested in the social aspects of the discovery process, especially as described by the American Pragmatist philosopher, G. H. Mead. (See in particular his "Scientific Method and Individual Thinker," in his *Selected Writings*, 1964.) There is a dynamic interplay, Mead says, between creative scientists and the groups within which they operate, but which also support them. Mead goes so far as to say that any epistemological account – and he discusses all those known to him at the time – that does not reflect the fact that creativity depends on communities of knowers is doomed to failure. In a famous phrase used later by sociologist of science Robert Merton, "We see further because we stand on the shoulders of giants." Yes, in science there are creative individuals, but their very creativity depends on their interaction with mentors. A little more reading in the American Pragmatists quickly revealed that this was no genius insight on

Mead's part. Beginning at least with C. S. Peirce and continuing with William James – with forebears all the way back to Descartes' own era among thinkers such as Giambattista Vico – the Pragmatists had recognized that Descartes' epistemological problem was a self-defeating pseudo-problem. Once we recognize that creativity is only fostered in groups, it becomes clear that fears on the part of individuals that they are being deceived by evil geniuses can themselves only gain traction in groups. All of us as graduate students in post-Cartesian settings may have perceived it as a real problem, but our fears could only be taken seriously by mentors who had taken their own doubts equally seriously – and promised us great rewards if we could "solve the problem." Even Descartes' own problem could only deserve to be taken seriously among a group of like-minded anti-Scholastics. And those with more serious problems on their minds – scientists, engineers, businessmen, ordinary citizens, you and I – cannot afford the luxury of universal doubt.

Once freed of the "epistemological problem," we are free to pursue serious discoveries. And it makes no difference whether the discovery is made within a so-called pure science community, or in applied or (what was then called) mission-oriented science, or in an engineering or technological community. So it was an easy step for me to turn my attention to technology, especially in the context of the late 1960s and early 1970s, when there were widespread critiques of the role of technology in the Vietnam War (we in the USA will recall John McDermott's famous article, "Technology: The Opiate of the Intellectuals," *New York Review of Books*, 31 July 1969), and in environmental issues (recall the first Earth Day in 1970). From that time on, for over 25 years, I devoted myself mainly to editing the books associated with the Society for Philosophy and Technology. I did publish a book of my own, *Social Responsibility in Science, Technology, and Medicine* (1992) – I'll talk about that next – and, as I will mention later, I ended up chronicling 30 years of controversies in SPT.

2. What does your work reveal about technology that other academics, citizens, or engineers typically fail to appreciate?

As I said, I linked my philosophical approach to that of the American Pragmatists, whom I read as linking their philosophical thought to activism in terms of helping to solve social or technosocial problems. (See Andrew Feffer, *The Chicago Pragmatists and American*

Progressivism, 1993.) I went so far, on one occasion, as to describe my approach as "social work philosophy of technology." (See Carl Mitcham, ed., *Research in Philosophy and Technology, vol. 16: Technology and Social Action*, 1997, pp. 3-14.) There, as elsewhere, I try to avoid the trap of being asked for a "respectable" defense of my approach. No "academic" defense of activism on the part of philosophers is needed, or called for. Ever since a Dewey-inspired (or at least Dewey-endorsed) definition of the academic's role in North American universities as involving a "service" component – not as an add-on but as essential to one's role as an academic professional – at least lip service has been paid to the demand that academics contribute to the good of society.

I believe that any separation between philosophy and the so-called "real" world, with its many problems, is not only arbitrary but pernicious. Dewey is famous for opposing all dualities, including that of separating philosophy from life, and I am in complete accord with him on that point. (See his *Reconstruction in Philosophy*, 2nd ed., 1948, and *Liberalism and Social Action*, 1935.)

I also agree with Dewey (and Mead) that, when philosophers get involved in social issues, they cannot play the role of philosopher kings, advising others how to solve their problems; they must get involved as equals with all those working on a particular issue, everyone from scientists and engineers to social scientists to government agents to experts of all sorts, but also including ordinary citizens involved in the issue. I have found this approach to be relatively rare among philosophers, but there are others who employ it; and in one of my books I appeal to technical professionals to do so in greater numbers than at present. (See my *Social Responsibility in Science, Technology, and Medicine*, 1992.) The professionals I appeal to there include technology educators, medical school reformers, media or communications professionals, bioengineers and biotechnologists, computer professionals, nuclear experts, and ecologists. In another book I appeal to fellow philosophers, especially fellow philosophers of technology and environmental philosophers, to do the same. (See my "Activist Philosophy of Technology: Essays 1989-1999," 2000, available on my University of Delaware website.)

In that set of essays, I argued for so-called applied ethicists – for example, engineering ethicists, or bioethicists, or environmental ethicists – to move beyond admittedly good academic contributions to the literature, to get involved in real-world activism to improve engineering professional societies and especially their policing efforts and lobbying efforts; to work with those struggling to make contemporary high-tech medicine more humane; and to actually apply environmental ethics theorizing to help improve the environment. (Andrew Light and Eric Katz seem to me to capture this best in their edited volume, *Environmental Pragmatism*, 1996.) It is in this last arena where I have chosen to do most of my activist philosophizing in recent years, as I have traveled repeatedly to Costa Rica to join with friends there in the ongoing effort to protect that country's amazing biodiversity, and especially the forests that are so important for that. Costa Rica is a model for the rest of the world, but conservation and preservation efforts there are ongoing, and continually threatened with backsliding. Echoing another philosopher who has done work in Costa Rica, David Crocker of the University of Maryland, I see my work as involving what he calls "insider-outsider cross-cultural communicators" (see the newsletter of the Institute for Philosophy and Public Policy, Summer 2004) – that is, non-resident outsiders with appropriate views and attitudes who work with like-minded people in a country or region to bring about change for the better.

3. What, if any, practical and/or social-political obligations follow from studying technology from a philosophical perspective?

I prefer not to talk about obligations so much as opportunities.

Our contemporary technological culture has many problems – probably no more than other cultures, but many – and I believe that the citizens of a democratic society have the right to expect that technical professionals, including philosophers, will contribute what they can to the solution of these problems. In my Social Responsibility, as I mentioned, I try to enlist biotechnologists and bioengineers – along with such others as ecologists – in such activism.

Presumably this question seeks an answer in the "ought" category, perhaps something like an ethical or social or even political obligation. But that's not what I think is called for here. The problems calling out for action in our troubled technological world are so urgent and so numerous – from global climate change to gang violence, from attacks on democracy to failures in education, from the global level to the local technosocial problems in your community – that it isn't necessary to talk about obligations, even social obligations. No, it's a matter of opportunities that beckon the technically trained – including philosophers and other academics – to work alongside those citizens already at work trying to solve the problems at hand.

Why? Can I offer a general answer to the question? I suppose I could try, but I don't feel the need to do so; certainly no urgency to do so. The problems are just there for all to see. And democratic societies have a right, in my opinion, to expect that experts will help them, experts from all parts of academia and all the professions. I would even go so far as to say that there is at least an implicit social contract (an ethical/social/political term that I won't define here) between professionals and the democratic societies in which they live and work and get paid for their professionalism. This may sound like rampant relativism: just get involved in any crusade you choose, as long as it "improves" society. To avoid this implication, I need again to fall back on American Pragmatism. It was the view of Dewey and Mead that there is at least one fundamental principle on which to take a stand: that improving society always means making democracy more widespread, more inclusive, inviting more groups – not fewer groups – into the public forum; elitism, "my group is better than your group," and all other such privilegings are anti-democratic. This "fundamental principle," however, is not just another academic ethics principle; it is inherent in the nature of democracy – at least as the American Pragmatists understood it. As I understand it.

I'm always happy when fellow philosophers try to provide academically respectable answers to questions of social obligation, of social contracts on the part of professionals, of the need to keep democracy open to ever wider inputs. But if we wait for them to provide such answers, it will typically be too late. Global warming proceeds apace. Loss of species diversity, of life on earth, proceeds apace. Threats to local communities in the so-called "developing world" in the face of economic globalization proceed apace. And so on and on. These and others like them are not issues of academicism. What I have in mind are urgent social issues that cry out for answers now.

I have been accused, on these grounds, of favoring activism over principle – even of abandoning the traditional role of philosophy as theoretical discourse. But I don't mean to do that. I believe Dewey was right in opposing all dualisms, including the dualism of principle versus practice or theory versus action. I welcome academic work on these issues; I just ask academics to accept activism as a legitimate part of philosophical professionalism. The issues seem to me that important.

Richard Rorty, who also calls himself a Pragmatist in the Deweyan tradition, shows some ambivalence on this issue: he talks about the death of philosophy, meaning mostly academic philosophy, but in one recent book, *Achieving Our Country: Leftist Thought in Twentieth-Century America* (1998), he has also lamented the failure of philosophers – and other academics – to involve themselves in progressive social causes, where his examples are relatively vague, such as "substituting social justice for individual freedom as our country's principal goal," though Rorty does endorse a claim that his favored "leftists" should agree on a "concrete political platform," on "specific reforms."

There is, in my opinion, much that philosophers can contribute. At the very least, they can contribute their "clear thinking" skills, but they can also make contributions out of the store of knowledge they have gained (if they have) from, for example, political thinkers going back through the centuries from Machiavelli all the way to the Stoics and Aristotle and Plato. And, as I have mentioned, bioethicists and environmental ethicists – among other so-called applied ethicists – can move beyond their academic work to help improve medical care or save planet earth. But I think it is going too far to talk about an obligation to do so.

4. If the history of ideas were to be narrated in such a way as to emphasize technological issues, how would that narrative differ from traditional accounts?

Don Ihde, in his *Philosophy of Technology: An Introduction* (1993) provides an account of the history of ideas in terms of the relations of human beings to their technologies. I don't think I can improve on Ihde's account. There was, of course, an earlier tradition among historians and social and cultural anthropologists that tried to link advances in social evolution to changes in the tools used by human societies (even some pre-hominid communities). That approach is now pretty much considered to be outdated, but it still makes sense to me to view history through the lens (Ihde likes that metaphor) of the changing types of technologies that various human societies have invented and utilized to improve their lives. Larry Hickman (see his *Philosophical Tools for a Technological Culture*, 2001, as well as his earlier, *John Dewey's Pragmatic Technology*, 1990) is now famous for arguing (he thinks that John Dewey first defended the view) that philosophy is a world-changing tool, a tool for improving society. In a related view, I have argued (see Durbin 1991, where I collect about a dozen essays by the most important philosophers of engineering writing at that time) that a philosophy of engineering is an important – and generally missing – part of philosophy of technology, in spite of Carl Mitcham's disparaging of "engineering philosophy of technology" (in his *Thinking through Technology*, 1994) as falling short of what philosophy of technology ought to be doing.

Sadly, one example often used by tools-oriented historians of technology is war-making technologies. (Neither Ihde nor I think they contribute much to society.) Others disagree, and some go so far as to say that the major advances in the history of technology have come from war-related technologies. So I'll grudgingly admit that it can be important to accept the bad with the good, in these terms, in understanding human history.

5. With respect to present and future inquiry, how can the most important philosophical problems concerning technology be identified and explored?

In a book MS, "Philosophy of Technology: In Search of Discourse Synthesis" that I recently completed – it should appear, online, in SPT's *Techne* (see spt.org) within a year – I try to do exactly this by way of a history of the major controversies within the Society for Philosophy and Technology. Others have done something similar; for example, Hans Achterhuis and colleagues do so in *American Philosophy of Technology* (2001), where a group of philosophers at the University of Twente in the Netherlands survey the work of a group of representative American philosophers, including many invited to contribute to this project.

In either case, the point is that giving philosophers of technology a fairer reading than they (we) often get in philosophical circles – indeed in the broader culture more generally – is an important way to get at such issues. In my MS on the history of controversies in SPT, I end with a dozen of these issues that I think make a contribution, whether to academic philosophy or to the improvement of society. I will list just a few here.

First, on the academic side, I believe Joseph Margolis's "constructivist" characterization of the knower and the known as inherently technological and pragmatic – ideas that he worked out in large part in contributions to SPT meetings – is a much better contribution to epistemology than the better known characterization of Mario Bunge. (See Margolis, *Reinventing Pragmatism:*

American Philosophy at the End of the Twentieth Century, 2002; and Bunge, *Treatise on Basic Philosophy. VII: Epistemology and Methodology III: Philosophy of Science and Technology. Part II. Life Science, Social Science and Technology.* 1985.)

Also somewhat academic is the issue whether or not philosophy of technology ought to have a place for traditional metaphysics. Carl Mitcham and Albert Borgmann and Fred Ferre have probably been the best known SPT philosophers who say yes; while Joe Pitt has been the most vehement arguer to the contrary. (See Mitcham, *Thinking through Technology: The Path from Engineering to Philosophy*, 1994; Borgmann, *Technology and the Character of Contemporary Life*, 1984; Ferre, *Philosophy of Technology*. 1988; and Pitt, *Thinking about Technology: Foundations of the Philosophy of Technology*, 2000.)

But most fundamental of all is the controversy within SPT publications over whether or not philosophy of technology ought to be an academic discipline in the first place. Recently, the issue seems to have been settled, with a resounding yes. (See Higgs, Light, and Strong, eds., *Technology and the Good Life?* 2000.) But I was not the only philosopher in SPT who argued, from the very beginning of the society, that philosophy ought to make a real-world contribution to the improvement of the world we live in.

Langdon Winner has been the most outspoken advocate of this approach, to issues through philosophers (or thinkers more generally). (See his discussion of social construction of technology work, in *Science, Technology & Human Values 18:3*, Summer 1993: 362-378.)

References

Achterhuis, Hans. 2001. *American Philosophy of Technology: The Empirical Turn.* Bloomington: Indiana University Press.

Borgmann, Albert. 1984. *Technology and the Character of Contemporary Life.* Chicago: University of Chicago Press.

Bunge, Mario. 1985. *Treatise on Basic Philosophy. VII: Epistemology and Methodology III: Philosophy of Science and Technology. Part II. Life Science, Social Science and Technology.* Dordrecht: Reidel

Crocker, David. 2004. "Cross-Cultural Criticism and Development Ethics." *Philosophy & Public Policy Quarterly, 24:3* (Summer): 2-8.

Dewey, John. 1920. *Reconstruction in Philosophy.* New York: Holt. Later editions 1948, 1957.

Dewey, John.1935. *Liberalism and Social Action.* New York: Putnam.

Durbin, Paul. 1968. *Logic and Scientific Inquiry.* Milwaukee: Bruce.

Durbin, Paul T., ed. 1991. *Critical Perspectives on Nonacademic Science and Engineering.* Bethlehem, PA: Lehigh University Press.

Durbin, Paul T., 1992. *Social Responsibility in Science, Technology, and Medicine.* Bethlehem, PA: Lehigh University Press.

Durbin, Paul T., 1997. "In Defense of a Social Work Philosophy of Technology," in Carl Mitcham, ed., *Research in Philosophy and Technology, vol. 16: Technology and Social Action.* Greenwich, CT: JAI Press. Pp. 3-14.

Durbin, Paul T., 2000. "Activist Philosophy of Technology: Essays 1989-1999."
Available on Durbin website,
www.udel.edu/Philosophy/pdurbin/Pub.html.

Durbin, Paul T.,Forthcoming. "Philosophy of Technology: In Search of Discourse Synthesis."

Feffer, Andrew. 1993. *The Chicago Pragmatists and American Progressivism.* Ithaca, NY: Cornell University Press.

Ferre, Frederick W. 1988. *Philosophy of Technology.* Englewood Cliffs, NJ: Prentice-Hall. Second edition, University of Georgia Press, 1995.

Hickman, Larry. 1990. *John Dewey's Pragmatic Technology.* Bloomington: Indiana University Press.

Hickman, Larry. 2001. *Philosophical Tools for a Technological Culture.* Bloomington: Indiana University Press.

Higgs, Eric; Andrew Light; and David Strong, eds. 2000. *Technology and the Good Life?* Chicago: University of Chicago Press.

Ihde, Don. 1993. *Philosophy of Technology: An Introduction.* New York: Paragon House.

Light, Andrew, and Eric Katz, eds. 1996. *Environmental Pragmatism.* New York: Routledge.

McDermott, John. 1969. "Technology: The Opiate of the Intellectuals," *New York Review of Books*, 31 July, pp. 25-35.

Margolis, Joseph. 2002. *Reinventing Pragmatism: American Philosophy at the End of the Twentieth Century.* Ithaca, NY: Cornell University Press.

Mead, G. H. 1964. "Scientific Method and Individual Thinker," in his *Selected Writings*, ed. A. Reck. Indianapolis: Bobbs-Merrill. Pp. 171-211.

Mitcham, Carl. 1994. *Thinking through Technology: The Path from Engineering to Philosophy.* Chicago: University of Chicago Press.

Mitcham, Carl, ed., 1997. *Research in Philosophy and Technology, vol. 16: Technology and Social Action.* Greenwich, CT: JAI Press.

Rorty, Richard. 1998. *Achieving Our Country: Leftist Thought in Twentieth-Century America.* Cambridge: Harvard University Press.

Winner, Langdon. 1993. "Upon Opening the Black Box and Finding It Empty: Social Constructivism and the Philosophy of Technology." *Science, Technology & Human Values 18:3* (Summer): 362-378.

6

Andrew Feenberg

Professor, Canada Research Chair in Philosophy of Technology

Simon Fraser University
Canada

Toward a Democratic Philosophy of Technology

1. Why were you initially drawn to philosophical issues concerning technology?

I grew up in the midst of the scientific world. My father was a nuclear physicist and although he declined to help develop the A-bomb during World War II, I was surrounded by people who did work on the bomb and whose children were my playmates. I recall the excitement of seeing the inside of scientific laboratories and even once a nuclear reactor. We were encouraged to make things with our hands and to enjoy learning about the universe. I am very much a child of the "atomic age" and so have never had that pseudo-aristocratic contempt for technology so common among humanist scholars.

Later, when I went to college, I got interested in technology issues through working briefly for the environmentalist, Barry Commoner, and through studying with Herbert Marcuse whose Marxist philosophy of technology was influenced by his teacher, Martin Heidegger. I was fascinated by their analysis of technology as Gestell or ideology, but skeptical of what appeared to be an ungrounded, transcendent critique. This was the background against which I wrote one of my first published articles, a study of dystopian themes in the European and American cinema of the 1960s.

In 1981, I was invited to work on the development of the first online educational program, at the Western Behavioral Sciences

Institute in La Jolla, California. This experience put me in touch with a developing technology. It soon became obvious that some of the principles of the classical philosophy of technology I had learned were false, or at least did not apply in this new field.

Up to that time computers were understood to be calculating and filing devices. We were part of a larger movement that transformed them into communication media as well. That movement originated in human agency and not in some prior essence of technology. It was hard to believe any longer in technological determinism while we ourselves were busy transforming the meaning of the computer.

Jacques Ellul had claimed that technocrats guided by the pursuit of pure efficiency were taking over, but this assumed that the technocrats knew what they were doing. Not so! This became clear to me the day a vice president of the second largest computer company in the world invited me to lunch and asked me to predict the future of computing. The absurdity of this high-level expert asking me, a student of Marcuse, for my opinion awakened me from my "dogmatic slumbers."

These experiences brought me to the realization that philosophy of technology should be my principal study. I wrote some of the early articles on online community and online education and began to rethink all my assumptions. I recall studying texts by philosophers of technology Gilbert Simondon, Langdon Winner and Albert Borgmann. I also found essays by Bruno Latour and several other sociologists of technology helpful.

At this time Marxism was increasingly questioned not only by conservatives but by Marxists too. I decided to write a book in which I would re-examine the Marxist approach to technology in terms of what I could still believe. This was the origin of the critical theory of technology that I continue to develop to this day.

I argued that the imposition of domination through the social shaping of technology was not confined to the factory. If Marx had originally conceived of the factory as the scene of decisive battles for control of technology, this was because only in the factory was technology bringing together vast collectives of relatively free individuals who could rebel with some prospect of success. But what happens when the whole of society is organized around large-scale technologies, when technical networks embracing thousands, even millions of people, extend into every area of social life? Then the factory is no longer privileged but becomes one of many scenes of struggle for control over technology, and by extension over society as a whole. This analysis offered a perspective on a wide and apparently disparate variety of new struggles in domains such as workplace health and safety, environmentalism, medicine, and education. With this insight I had the makings of a theory of my own.

2. What does your work reveal about technology that other academics, citizens, or engineers typically fail to appreciate?

Academics, citizens, and engineers each have their blind spots as no doubt do philosophers of technology. However, philosophers of technology have the advantage of taking into account the whole field, including the views of these other groups. They note for example that academics tend to ignore technology, to abstract from it in the constitution of objects such as society and politics. But what sense does it make to talk about social institutions or the political order without situating them within the technical context that mediates the human relations in which they consist? Citizens seem inclined to overestimate technology, demonizing it or worshiping it depending on the latest news. In both cases technology appears as an independent power external to society. But the history of technology reveals social influences everywhere in its development. Engineers have a more realistic view of technology but often seem uncritical where criticism is merited, for example, in the case of nuclear power. One does not have to be a technophobe to mistrust specious reassurance about dangerous technologies.

The most important thing we have learned in philosophy of technology is the ubiquity of its influence on our lives. It is said that the fish do not know that they are wet. Like fish we swim in a sea of technology without noticing its presence except of course when things break down. But in taking technology for granted as a kind of natural milieu we lose sight of its contingency on our own choices and actions. Philosophy of technology can make us aware of the extent to which technology forms the environment in which we live and, correspondingly, the extent to which we have the power to change this environment.

3. What, if any, practical and/or social-political obligations follow from studying technology from a philosophical perspective?

The main obligation philosophy of technology teaches is responsibility for our own creations and for the consequences of our own actions. We know we should take such responsibility in personal affairs, but what about our relation to nature and to society? Most of our institutions and received ideas tell us the natural world is a vast grab bag and garbage dump for which we have no responsibility at all. As for society, we are told that our responsibilities begin and end with paying taxes and voting. These are catastrophic errors. Technology is a collective project of society as a whole and can only be brought within the scope of our ethical obligations through a wide variety of political interventions, including protests, boycotts, and active collaboration with experts around new visions of the technical future.

This is why I am most concerned with the implications of technology for democracy, a subject that is still largely overlooked. Technologies form the framework of our lives but they are designed with little or no democratic input. This is a serious failure of our institutions. It must be addressed by reforms in education, the media, the corporations, law, and the technical professions.

The idea of democratizing technology has many sources. Perhaps the two most important philosophers to advocate this idea were Marx and Dewey. Marx believed that worker control of the factory could transform modern society and the technology on which it is based. Dewey also hoped for wider citizen participation in technological decision-making. Neither had significant examples of democratization to point to. Furthermore, technological determinism was far more popular than their democratic position until quite recently. Indeed, Marx was understood as a determinist for generations.

In recent years this has begun to change due to democratic interventions into technology by users and victims and frequent calls for alternative technologies from scientists and technical experts in fields such as environmental protection and medicine. It is time to develop a democratic philosophy of technology to explain and further this important change in the very idea of politics.

This orientation is common to many of us in American philosophy of technology and distinguishes our work from specialized investigations of particular technologies and technical systems. In some cases philosophers of technology are involved in practical applications of their ideas.

In my own case I felt called to do something at the end of the 1990s when online education became a subject of public discussion. Although online education was invented by faculty in the early 1980s as a way of placing old-fashioned seminars online, at this time university administrations, computer companies and so-called futurologists were all touting the delivery of automated course materials on the Internet as a solution to budget problems in higher education. With the automation of education on the agenda, employing the very technology I had pioneered, I felt responsible for setting the record straight. I made a number of speeches calling for faculty control of online education, and contributed to a movement among faculty organizations to impose a more traditional pedagogy on the new technology.

In the course of these debates I realized that the software being made available by the advocates of automation was ill adapted to the discussion centered pedagogy I advocated. I designed alternative software and implemented it under a grant from the US Department of Education. I am still working on this "TextWeaver" software.

I consider this a small example of a democratic intervention into technology. The introduction of human communication on computer networks, of which our efforts in education were a part, has had a considerable impact. Other important interventions have occurred around environmental issues and such medical issues as childbirth practices and human experimentation on AIDS patients. Although the conventional political process is ever more discouraging, technical politics offers a ray of hope.

4. If the history of ideas were to be narrated in such a way as to emphasize technological issues, how would that narrative differ from traditional accounts?

In fact the whole structure of the history of ideas would be challenged. I have written about this in my book on Heidegger and Marcuse. I argue that the model of technical making has shaped the fundamental categories of Western philosophy, existence and essence, nature and culture, and so on. I believe craft is the origin and model of these categories which philosophers in antiquity extended to nature as a whole. We still live with the consequences even as this original foundation breaks down.

Consider for example the concept of essence. This was a technical term in ancient Greek philosophy. Plato introduced it as the *eidos*, or "ideal form" as it is commonly translated. The *eidos* contains the essential meaning of a class of things and exists independently of those things. But this exactly describes the relation of the craftsman's project to the thing he creates. The project is present in his mind as a specific meaning before he installs it in reality in an artifact. Furthermore the project has a sort of objectivity in a society where craft is regulated very strictly by tradition and custom. But natural objects do not depend on a project and so the separation of their essence from their concrete existence is an *aposteriori* act of the human mind.

Yet from Plato until Hegel and Nietzsche Western philosophy lived in denial of this difference and assumed that the natural world was structured very much like the world of artifacts. In Aristotle, whose thought shaped the tradition for millennia, the origin of essence is attributed not to the human mind but to things themselves which are supposed to have an inner principle that governs their form and development.

The impetus for a philosophical critique of the concept of essence came from natural science which from Galileo on rejected the Aristotelian understanding of nature. However, without a concept of objective essences, the status of meaning in general was thrown into question. Two main solutions were offered. For Hegel (and following him for Marx) meaning is a product of historical development and relative to culture. For Kant and Nietzsche individual reason or will constitutes meaning. In the 20th century the negative critique of meaning implied in these solutions is incorporated into a technological framework: meaning is a tool used by state and corporate organizations to define the environment and manipulate those who live within it. Essence, which emerged from craft, is now a technological plan.

One of the great dilemmas of philosophy since the early 20th century is how to take account of the breakdown of Aristotelian substantialism without falling into this technological positivism. If things are not solid unities determined by an inner essence, then are they mere assemblages of meaningless materials manipulated in accordance with human intentions, or more precisely, with the intentions of the powerful? But even human intentions dissolve if human beings are as contingent and disunified as the things they manipulate. What is it that holds humans together as apparent substances and gives us the impression that society is a solid "thing" capable of constituting a world?

In the face of these paradoxes, Lukács, in 1923, identified "dereification" as the fundamental task of philosophy. The task has been taken up again and again in relation to science and technology from Dewey and Heidegger to Foucault and Latour. We all pursue this task on the edge of a precipice below which lies absolute nihilism. Too much emphasis on the process of assemblage and the human subject disappears in behaviorist mechanism. From that standpoint resistance and democracy are meaningless. Too much emphasis on human agency and the theory risks falling back into a substantialism of the subject, an idealism without norms or limits of any kind. The unbound Prometheus beckons on this path, but precisely what we need today is a limit on technological hubris.

5. With respect to present and future inquiry, how can the most important philosophical problems concerning technology be identified and explored?

This perspective on the history of ideas leads us to a very radical understanding of the present state of modernity. The nihilistic world of modern organization finds no limits on its destructive powers. Somehow humanity must find resources for establishing such limits. It is interesting to note that the concept of limit, or *peras*, is closely related in Greek thought to the concept of essence. The limit of the thing is the boundary established by its essence against the chaos of meaningless matter. The thing is forged into its essence, most obviously by the potter or sculptor who shape a shapeless mass in accordance with a meaning. We moderns must find an equivalent of this Greek *peras* without returning to a dogmatic concept of essence.

How to do this? One hint: philosophers should follow the news a bit more carefully and come down from their conceptual heaven. The world is undergoing many drastic changes and these are reflected in public enthusiasms, concerns and protests. Philosophers should pay attention and discover within this turmoil the emergence of new limits, new meanings or norms, which increasingly have the power to intervene in the shaping of modern technology and the society that depends on it.

This is a loosely Hegelian approach. Norms can no longer descend on us from a transcendent heaven. We can no longer hear the voice of the gods, and abstract "reason," their modern stand in, has lost its charms. But an arbitrarily constructed norm has no force, is hardly worthy of consideration as a norm, indeed is merely an expression of the will to power. If we cannot rely on the eternal and unchanging commands of gods or reason, we must find value emerging in history in the course of struggles to impose a humane way of life on the resistant material of inherited traditions and institutions. We must be able to say "All men are created equal," and if we can no longer justify this proposition with reference to God's creation, then we must do so on the basis of the evidence of history which tends to prove the proposition in practice for one human group after another and perhaps someday for humans in their technical involvements as well. This is the deepest philosophical significance of the democratic movement that has unfolded since the 18th century.

This is not to say that we can be confident of success in controlling technology democratically. But philosophy can argue from the evidence history provides to a general concept of a redeemed modernity regardless of the likelihood of success. That concept can contribute to its own realization by raising people's expectations and hopes and giving a larger context to their protests and creativity. This is as much as one can ask from philosophy and it is by no means an insignificant contribution.

Books

Critical Theory of Technology, Oxford University Press, 1991.

Technology and the Politics of Knowledge, co-edited with A. Hannay, Indiana University Press, 1995.

Alternative Modernity: The Technical Turn in Philosophy and Social Theory, University of California Press, 1995.

Questioning Technology, Routledge, 1999.

Transforming Technology, Second edition of Critical Theory of Technology, Oxford University Press, 2002.

Modernity and Technology, co-edited with Tom Misa and Philip Breij, MIT Press, 2003.

Community in the Digital Age, co-edited with Darin Barney, Rowman and Littlefield, 2004.

Heidegger and Marcuse: The Catastrophe and Redemption of History, Routledge, 2005.

See also, http://www.sfu.ca/~andrewf/

7

Joan H. Fujimura

Professor

Department of Sociology and
Holtz Center for Science and Technology Studies
University of Wisconsin-Madison
USA

---

1. Why were you initially drawn to philosophical issues concerning technology?

I grew up in Hawaii and had hiked through the ethereally beautiful Alakai Swamp with my biology teacher who lectured us on myriad species of plants and birds that existed nowhere else on this planet. In high school, I liked math and science classes and spent a summer as an intern in a laboratory at the University of Hawaii-Manoa. In college, I began pre-med biology courses, but soon found that I preferred studying knowledge through philosophy and the social sciences. We read Kuhn's *Structure of Scientific Revolutions*, which immediately entranced me. I had found my home and my passion: how is knowledge produced, under what conditions, through which processes, who are the actors. But this passion is a result of more than an interest in the production of scientific knowledge. It is a result of growing up in multi-ethnic Hawaii where you can say something in one language that simply has no words in another language. It is a result of moving from Hawaii to the "mainland," where I found a world that was full of different ways of doing things and wanted to understand from those differences. I was entranced with different ways of organizing social and natural knowledges. Further, I saw these multiple ways doing things as different but not inherently hierarchically organized. To my mind, hierarchies of knowledge were (and are) social productions.

In graduate school, my interest in the production of knowledge became more oppositional, perhaps a sign of my generation.

Opposition 1: As a graduate student in Sociology at University of California-Berkeley, I was studying race and gender, after having spent time in Sweden and Japan studying these countries social welfare systems. I was amazed by the different positions of women in these countries and in the U.S. and had contemplated a study of these contrasts. However, I also noticed during my graduate study that race and gender were considered lower status subject areas in the discipline. I also was studying sociology of work and occupations, which had taught me that all specialty areas of science and medicine were historical productions and heavily embattled productions at that. Thus, one oppositional impetus was to show my department (and the discipline) that the status hierarchy of areas of study within sociology was socio-political and historical productions, that there were no inherent "intellectual scales" by which they could justify their value determinations. It was also then that I added another question to my list of "who, what, when, where, how": What are the stakes?

Opposition 2: Another opposition came by way of Lucien LeCam, a statistics professor, who was working with a medical physician Alan Levin and his wife and fellow researcher Vera Byers at University of California San Francisco on a study of transfer factor in cancer research. Transfer factor is a substance prepared from human white blood cells of healthy donors close to the patient. It is used to attack the patient's cancer cells. In the late 1970s, Byers, Levin, and LeCam showed that six of seven patients treated by Levin and Byers were disease free 62 to 82 months after treatment. However, they had difficulty getting funding agencies and other researchers to treat seriously transfer factor as a potential therapy for cancer. When the Mayo Clinic decided to evaluate transfer factor, they did so using a quantity that was much smaller than that used in the original study and Mayo's study did not yield results similar to those of the Levin-Byers-LeCam study. This case intrigued me. These were legitimate scientists, not snake oil salespeople. Yet, they had differences with the Mayo Clinic's study.

I decided then to study cancer research as my dissertation topic. It gave me the opportunity to study epistemology in action, to show how knowledge production is a contextual and situated set of practices. It also afforded me a wide terrain for studying different approaches to the study of cancer. After some reading, I decided to focus on genetic versus epigenetic approaches to explaining the causes of cancer because I had interviewed people on both sides at

U.C. Berkeley and so thought I had an example of a controversy. I had also been reading late 1970s and early 1980s science studies writers who focused on scientific controversies as pathways to the inner workings of science production. After a month, I realized that the controversy existed between only a few people. Instead, most people were already entranced by oncogene and other genetic approaches to cancer causation. So I then proceeded to try to understand why: that is, who, what, when, where, how had cancer causation become a genetic problem? I ended up then studying how cancer genetics became normal science.

To say all this more succinctly, I added to my past interest in the diversity of knowledges the idea of knowledge as power and hierarchy. I wanted to understand how that power was constructed and produced.

2. What does your work reveal about technology that other academics, citizens, or engineers typically fail to appreciate?

The research I did earlier in my career focused on the sociopolitical production of technoscience. This in itself was not new. French sociologists and historians, American feminists, historians of eugenics, Fleck, and Kuhn, among others, had previously written on this topic. (This list fails to mention writers in other countries whose names I am still learning in my current explorations of postcolonial studies of science.) In contrast to these earlier writers who focused on thought styles, paradigms, intellectual racism and sexism, my work focused on the practices and processes of scientific work. It showed that we could study scientific research as work. To emphasize this, for example, one of my chapter titles was "Problems and Work Practices: Improving on the Shop Floor" (Fujimura 1996, Chapter 6). This is opposed to many philosophers or sociologists who saw knowledge production as something different (read "special") from other kinds of work. It is different in some ways, but we can also learn from thinking in terms of its similarities to other kinds of work. That is, by examining research as ordinary work performed by ordinary actors, we can see things that we do not ordinarily see. For example, like other kinds of work, scientific research today most often takes place in institutions (universities, for-profit research institutes, pharmaceutical companies) that reward people for their productivity—at least in capitalist economies. Thus, it was important for my research to examine how the organizational and reward structures of technoscientific work affected the content of produced knowledge (and not only priority disputes as in Merton's work). In order to see any effects, I conducted a close ethnographic study of the practices within laboratories. In the case I examined, I argued, in brief, that some researchers produced a new technology (a DNA probe) together with a theoretical explanation for what this probe did, and proceeded to sell their frame (I called this a theory-methods package) to other researchers in many subfields of biology. These other researchers used this technology plus theory to revamp old, worn projects—to retrofit their laboratories to produce new knowledge. Given their location within an institutional structure that rewarded production and productivity, these scientists were impelled to constantly produce new knowledge and, in the 1960s and 1970s, within shrinking time frames. (This has its own history, but waits for another paper.) When older projects failed to produce new answers in short time frames, these scientists were in some sense forced to look elsewhere, to attempt transitions to new research. Similarly, younger researchers also needed to think in terms of projects that could produce publishable and fundable results within short time frames. That is, researchers turned to projects that could help them construct "doable" problems.

I showed then that new knowledge (in that case, oncogenes or cancer causing genes) gained their existence in part because of this reward structure (and other factors, of course). That is, the reward structure framed the conditions under which a lot of people chose to conduct research on oncogenes and to jump on the oncogene bandwagon. (That did not mean that oncogenes are not "real;" things are real in their consequences. Indeed, oncogene research led to the production of many new lines of research in biology.) In this case, my comparative analysis of work showed that organizational and institutional opportunities and constraints helped to shape the very content of knowledge. Another kind of knowledge could have been the result. Note, however, that this does not translate into organizational or institutional determinism. In contrast, my work examined knowledge production in a particular place and time, within a context with particular historical and institutional features. These features too could have been otherwise. Had the particular historical and institutional features been different, they may have produced a different kind of knowledge. Nevertheless, organizational and institutional work practices had a hand in producing oncogenes as cancer causing genes beginning in the mid- to late- 1970s.

The above study showed that all knowledge is contextual, is situated, and should be studied in its context of production. Some knowledges are more obviously influenced by socio-political factors than others (as I will discuss below), but all knowledge is situated.

3. What, if any, practical and/or social-political obligations follow from studying technology from a philosophical perspective?

In contrast to my example above, my current work focuses on more transparent forms of socio-political and cultural production of scientific knowledge and explicitly addresses practical and socio-political outcomes. For example, one of my new projects examines issues of race, biomedical genomics, and disease. The question I ask is: can biomedical researchers incorporate race as one of their predictor variables without reifying race as a genetic category?

Anthropologists and biologists have long argued that race categories as proposed by early 20th century anthropologists do not constitute genetically distinct populations and are instead socio-historical categories that differ by locality. Building on these earlier studies, sociologists, anthropologists, historians, and philosophers have discussed the social construction of race classifications and categories and the diversity of race classification schemes across societies.

In the last five years, however, a debate has been raging in biological, medical, and public arenas about race and biology. Some have argued that race is useful in medical diagnosis and treatment and that there are biological and even genetic differences between different racial and ethnic groups. Others respond that race as a social designator may sometimes be useful as a diagnostic category because differential experiences of racism may lead to lower socio-economic status and thereby greater exposure to stress, toxic waste, poor nutrition, which together contribute to differential health outcomes. However, they argue, race is not useful for therapeutic research, since a therapeutic correlation assumes that social race is equivalent to a biological population. These writers argue against recent FDA approval of the hypertension drug BiDil for use specifically in African Americans and research on the drug Iressa which its makers argue is particularly effective in Asian Americans. Still others argue that the more dangerous effect of human genetic variation research and pharmaceutical research on race-specific treatments is their potential to reinstitute biological categories of race. By now, Francis Collins, Director of NIH's National Human Genome Research Institute, is aware of the debate and the potentially dangerous interpretation of contemporary human genetic research and has asked bioethicists for advice on how to avoid such interpretations from occurring.

However, is the problem just one of interpretations made from the results of human genetic studies? Are there social and medical consequences of incorporating race as one of their predictor variables, despite the careful attention to interpretation? Is it feasible to use race as a variable in human genetic studies and still avoid reification?

To explore such questions, I am engaged in a project with my co-PI Pilar Ossorio that will examine the conceptualization of "population" in the organization of the collection and analysis of the data in several human genetics studies. While sociologists, anthropologists, and bioethicists are currently studying the potential consequences of human genetic variation studies (e.g. NIH's haplotype mapping project) on racial/ethnic communities, none has focused on the source of the problem: that is, the notion of population as operationalized in the practices of human genetic studies.

Our project examines how notions of population are conceived and operationalized in basic research laboratories that study human genetic variation. By focusing on the design, data collection, and analysis processes, this project will produce a description of the process by which social designations of categories become transformed into genetic populations through the detailed and quotidian decision-making processes of research.

Our preliminary studies show that there is no consensus on how to define "population," and there are no standard practices for operationalizing the term "population" in human genetic variation studies. Instead, terms like race, ethnicity, or national origin are often used as heuristic devices in the process of collecting biological materials. These materials are then analyzed for their genetic composition in the project's research process. At this point, the sample categories often become translated into "populations," where the meaning of population has shifted from the original heuristic to a genetic meaning. If the information generated through this process is then used to organize "common heritable variations" along particular categories, this presents both practical (medical) and political problems. The latter, for example, may conflate heuristic categories with "biologically distinct populations."

In our initial studies of the literature, we found widespread ad hoc use of socio-cultural notions of human groups such as race, ethnicity, language group, and nation in human genetic studies. Thus, local, informal notions and categories of difference and similarity in various societies (e.g., the United States, Japan, the African continent, China, the European continent) come to play significant roles in the operationalization of the concept of population in data collection and analysis. Further, there is much slippage in the way the term "population" is currently used in genetic variation studies. The term is currently used to refer to everything from a collection of people in sites around the world or from the US (such as the "White" population) to a "Mendelian" population. That is, they treat race as a set of biological categories that are bigger versions of earlier, smaller geographically isolated groups.

Our study of the processes that bring race and ethnicity into the arena of genetics and medical research will contribute to several areas of research, including the sociology of race and race classification which is of interest to demography and population studies; to the study of categories and classification more generally; and to the social studies of science. It will also contribute to bioethics, anthropology, human genetics and human genetic variation, biomedicine, and population health. We hope that these contributions will together be able to then influence medical professionals, the American Medical Association, insurance companies, and government policies to make certain that their practices do not use race (and other social or folk categories) in ways that can be reified as biological or genetic categories.

4. If the history of ideas were to be narrated in such a way as to emphasize technological issues, how would that narrative differ from traditional accounts?

This is a broad question, so I will interpret and answer it in three ways.

I first interpret it as a question about the contributions of recent social studies of science to pre-1980s philosophy of science. Most philosophy of science of that earlier era tended not to examine actual experimental practices and technologies in laboratories, whether the laboratory was in a building or out in the world (e.g. population genetics fieldsites, ecological fieldsites, pharmaceutical companies, and patient activist groups). Technoscience studies focused on practices in situ through ethnographic studies in laboratories and other sites of production. These studies added observations and analyses of the practical and material practices of the production of knowledge to the previous discussions of ideas and representations. These observations and analyses differed depending on the particular theoretical and methodological approach of the science studies analyst, whether it was ethnomethodology, symbolic interactionism, actor-network theory, SSK, anthropological ethnography, etc. Nevertheless, this study of the actual practical and material production of knowledge was new.

There were critiques of these ethnographic studies. For example, some scholars raised questions about whether we were reflexive about our positions as observers of action. That is, how were our observations "better" than those of the scientists we studied. Other scholars also raised another criticism: In our ethnographies of material practices and production, were we not just relying on the scientist's representation of the object? I believe that the answer to those questions is relatively simple. We ethnographers added one more perspective on the actions of the material agents, and we also added our observations of scientists' actions. We reported about scientists' interactions with their materials and technologies, with other scientists, with their funding agencies, with their audiences, and more. Some of our observations differed from the observations of the scientists conducting experiments. These kinds of interactions and observations were often absent from traditional histories and philosophies of science. Like James Clifford and George Marcus' (1986) analysis of anthropological ethnography as contingent historical, and contestable, one could represent us as just one more layer of observers. Our work, like that of other anthropological ethnographers, could be considered to be relative knowledge. However, I prefer to take the position of feminists like Donna Haraway who argue that we present different and compelling views from elsewhere.

A second interpretation of the question: what happens when we add the study of natural objects to the study of ideas? Traditional accounts of the history of ideas also tie into traditional accounts in the sociology of knowledge. For example, Mannheim (1936) assumed that sociologists could study the social production of social knowledge, but not natural knowledge. For the actions of nature, he assumed that we would have to rely on natural scientists. Most sociologists have generally ignored the natural sciences, except for those who chose to criticize it from the outside and those who chose to uncritically accept scientific accounts as the authoritative knowledge upon which sociology is to be built. In contrast, I argue that feminist studies and the social sciences can engage with the materiality of nature and become activists in the production of biological knowledge.

This is what the social and feminist studies of science have contributed to the social sciences over the last thirty years. We have argued that humans and nonhumans make the world together, therefore it is our obligation to study this interactive production of knowledge and technologies [see, for example, Haraway (1991) on material-semiotic actors and Callon and Latour (1992) on nonhuman actors]. These studies of the production of natural knowledge have been performed in three different ways. First, we have examined how humans read our social understandings onto nature and therefore how humans use nature to justify social organization. Second, the study of technoscience has taught us how scientists can and often do produce knowledge that challenges the ways in which people view the world. Third, our studies of technoscience can engage us directly in the production of knowledge. That is, we can deconstruct scientific research to find data (what I call "awkward surplus") that have been ignored or left unexplained by scientists, or we can work directly with scientists – collaboratively or oppositionally – to interpret the actions of nonhumans (Fujimura 2006). That is, we do not have to leave the natural sciences entirely to the natural scientists.

However, I do not mean that social scientists should collect DNA samples and use them to make knowledge about human behavior, as some social scientists are proposing. Most social scientists do not have years of training necessary to understand the complexities of genetic actions. Instead, I argue that we can contribute our knowledge about social organization, social psychology, socialization, and more to understand how scientists bring their own understandings of society to their work on nature. In my study of sex determining genes, I first show how molecular genetic researchers read the social organization of gender binaries onto their research on sex determining genes (Fujimura 2006). However, I also show how my study of their research yields information about genetic complexities and variations that can be used to explain variations among sex phenotypes. Molecular geneticists performed the experiments that yielded this information, but they ignored it. My study highlights this awkward surplus and its potential to explain variations in sex. I hope that these findings will also translate into tolerance for such complexity and diversity among humans. As an example from another field beside genetics, Lucy Suchman (2006) has examined how artificial intelligence researchers conceptualize intelligence and cognition. She finds that when AI researchers attempt to solve a problem themselves—in one case, how to work a photocopy machine—they do not cogitate the way that they think they or other humans do. When Suchman examines how these AI researchers cogitate, she finds a variety of patterns that could then be used to produce far more interesting theories of cognition.

Related to this last example is a third interpretation of Question 4: How have new technologies and knowledges transformed society? In contrast to the history of ideas, science and technology studies have, for example, studied the relations between physics, bombs, and war in the 1950s and 1960s. Feminist scholars of technoscience have examined the changes wrought by the introduction, for example, of refrigeration; it changed the time women spent on preserving food, but did not change the time spent on housework generally. In the 1970s, scholars discussed the potential consequences of recombinant DNA technology. More recently, other science studies scholars have discussed the simultaneous production of science and society through the production of new biotechnologies, genomics, stem cells, and nanotechnologies. The more interesting of these studies discuss the role of technoscience not as simple technological determinism, but as the complex co-production of science and society.

5. With respect to present and future inquiry, how can the most important philosophical problems concerning technology be identified and explored?

I have four short answers to this question. As a graduate student in history, historian and labor activist Howard Zinn realized that he had never read about the labor movements and protests that he as a youth had witnessed in the streets of New York City. They were missing from historical accounts. So he proceeded to study and write about them. So I think we should ask what Zinn asked: What is missing? Second, Donna Haraway's (1988) article on situated knowledges still inspires me many years after I first read it. She argues there that knowledge makers should include those who have stakes in that knowledge, especially including those who live with the consequences of the knowledge and those who are objects of knowledge-making. From this, I take our job to be the examination of what are the stakes and who are the stakeholders in each situation where knowledge is produced or applied. Third, transnational science is one of my current interests (Fujimura 2000, 2003). This includes postcolonial studies of science and global studies of science. I prefer here to focus my attention on the located sites, á la Haraway's pointers, since that is the primary way to learn who is affected and how. But I also attempt to study connections between local sites or what I call transnational science. This is difficult to accomplish without acting like the universal knower of Euro-American science and therefore requires careful and judicious attention. Fourth, despite what I have learned about technoscience gravitating towards doable problems, I believe that we should study complexity complexly (Fujimura 1996, 2000, 2003, 2005, 2006). Indeed, as is evident in my answers to the five questions, three themes that permeate my work are complexity, variation, and diversity. These three are key words for identifying and exploring the most important philosophical problems concerning technoscience.

References

Callon, Michel and Bruno Latour. 1992. "Don't throw the baby out with the Bath school!" In Andrew Pickering, ed., *Science as Practice and Culture.* Chicago: University of Chicago Press.

Clifford, James and George E. Marcus (eds.). 1986. *Writing Culture: The Poetics and Politics of Ethnography.* U of California Press.

Fujimura, Joan H. 1996. *Crafting Science: A Socio-History of the Quest for the Genetics of Cancer.* Cambridge, MA: Harvard University Press.

—2000. "Transnational Genomics in Japan — Transgressing the Boundary Between the 'Modern/West' and the 'Pre-Modern/East.'" In *Doing Science + Culture.* Roddey Reid and Sharon Traweek, eds. New York and London: Routledge Press, pp. 71-92.

—2003. "Future Imaginaries: Genome Scientists as Socio-Cultural Entrepreneurs." In A. Goodman, D. Heath, S. Lindee (eds.), *Genetic Nature/Culture: Anthropology and Science Beyond the Two Culture Divide.* Berkeley: University of California Press, pp. 176-199.

—2005. "Postgenomic Futures: Translations Across the Machine-Nature Border in Systems Biology," *New Genetics and Society, vol. 24, 2*: 195-225.

—2006. "Sex Genes: A Critical Sociomaterial Approach to the Politics and Molecular Genetics of Sex Determination," *Signs (vol. 32, Autumn, no. 1)*: pages.

Haraway, Donna J. 1991. "The Biopolitics of Postmodern Bodies: Constitutions of Self in Immune System Discourse." In Haraway, *Simians, Cyborgs, and Women: The Reinvention of Nature*. New York: Routledge, pp. 203-230.

—1988. "Situated Knowledges: The Science Question in Feminism as a Site of Discourse on the Privilege of Partial Perspective." *Feminist Studies* 14(3): 575-99.

Mannheim, Karl. 1936. *Ideology and Utopia*. London: Routledge & Kegan Paul.

Suchman, Lucy. 2006. *Human-Machine Reconfigurations: Plans and Situated Actions*. Cambridge University Press, 2nd edition.

8

Peter Galison

Pellegrino University Professor

Harvard University
USA

---

1. Why were you initially drawn to philosophical issues concerning technology?

I idolized my great-grandfather—an old-school electrical engineer who had worked for Edison's laboratories in Menlo Park, New Jersey, and who continued to build and patent high-voltage devices into his 90s. Late in his life he had a basement lab, something Dr. Frankenstein would have loved, jammed with wondrous things: a lathe that made any kind of screw you wanted; shelves of glass parts he had blown; heaps of copper coils; bottles of mercury that I found particularly entertaining (as a result of which I probably conduct electricity rather well even today); and giant double-pole, double-throw switches that launched sparks from gun-shaped probes. All this meant I grew up surrounded by a splendid array of bakelite-mounted meters, batteries, and half-disassembled radios. I tried running an amateur radio station from my bedroom, but managed mainly to send parasitic signals into my New York City neighbors' TV's. I don't think I've ever lost a deep-running sense of the other-wordliness of these technical objects.

I actually didn't much like what I learned early on in history; what finally grabbed me was history grounded in material things—the Annales school historians, especially Fernand Braudel and Marc Bloch, appealed to me because they wanted to know how long a letter would take to cross from Paris to Rome, or how the farming methods fixed the ownership and policing of land. Marx interested me not so much for notions of ideology, but for his insistence that a philosophy of history could never be understood without grappling with the cubic meters of space each loom operator had in which to live. I liked—still do—his extraordinary insights into the shifting forms of labor that entered with differing technical structures. Out of the specific forms of work (for example, cottage-based "home work" or centralized steam-driven factories), he went on to analyze the history of labor relations and value.

I spent a year in Paris back in 1972—I was 17—working in a plasma physics laboratory at the old, still-in-the-fifth arrondissement Ecole Polytechnique. It was startling that these table-top devices—I worked on a cylindrical glass and steel device known as Q-machine–could probe the dynamics of plasmas. I loved the idea that abstractions—Boltzmann's transport equations and the like—had something to do with the delicate curves traced on graph paper output. It seemed like a miracle to me then and still does that lock-in amplifiers and copper screens could measure the bunching of electrons as an electrical wave passed through the gas. In a way, the major part of my work still addresses that sense of astonishment: How do machine-based demonstrations sync with reason with theoretical equations, thought experiments, and rules of thumb? How does conviction in its various forms, take root—here in a set of equations, there in the midst of machines?

2. What does your work reveal about technology that other academics, citizens, or engineers typically fail to appreciate?

There is an expectation that the river of technical development flows in one way only: from the springs of abstract science, through canyons of applied science, and eventually out to the delta of technology. Much of my work aims at studying the reverse flow—the myriad ways in which technology not only enters through the utility of specific devices, but also conditions ways of thinking, kinds of evidence, formation of concepts. *How Experiments End* (1987) is about the distinct subculture of experimentalists, and how, in particular, experimentalists (in ways linked to but distinct from theorists) find their way to say: "this particle or effect exists and is not an artifact of the apparatus or environment." This coming-to-closure is not a debased theorem, but its own kind of argument. The theorist might say: anti-protons have to exist, their existence follows inexorably from the existence of protons and the theory of special relativity. For the experimentalist such claims were fine—but the conviction that anti-protons really were there could only come from the clicks of an array of electronic counters or the darkened grains of silver halide in a sheet of nuclear emulsion.

Image and Logic (1997) began where *How Experiments End* stopped—in the epistemology of certain machines, machines that make images on one side, and machines that produce logically assembled counts on the other. These traditions, these instrument subcultures, are *not* reducible to the strategies deployed by theorists, and lie quite often distinct from any particular experimental problem. This diversity within the practices of physics (experimenting, theorizing, instrument making) can therefore be pushed even further, into a distinction between instrument making traditions of image and logic. Theorizing too splinters on closer view into characteristically algebraic and characteristically geometrical-visualizable forms.

There are several historical-philosophical advantages to a more differentiated analysis of technical practices, one that joins procedures and arguments. First, it takes us away from the view that splits science, synchronically and diachronically, according to what objects are taken as foundational. National Bureau of Standard engineers in the early 1950s were building thermonuclear weapons in the South Pacific; Berkeley physicists were pursuing mesons—but there was a steady exchange between them because they shared an enormous array of concerns in the production and manipulation of liquid hydrogen. An over-focus on ontology blocks instead of facilitating an understanding of the movement of technical skills, procedures, objects, and people. Second, the view that science divides into blocks, with its insistence on "programmes" and "paradigms" shuts the windows and slams the doors between historical periods, and by so doing fails to capture the felt sense of continuity among working scientists and engineers. Accelerator-based physicists at Stanford in the postwar years may have started with a notion of the proton or neutron as elementary objects; using the same sort of machines and analyses they moved first to a view that there was some kind of structure inside these objects, and later to the notion that there were hard, point-scattering particles, and eventually to the view that these were dynamic quarks. Work continuity—continuity at the level of practices—could trump shifts in the list of what there is.

But if a more precise notion of scientific practices helps with the problem of felt continuity across scientific ontologies, it creates a new problem—perhaps even worse: *how* does communication work if each scientific culture has its own zoo of entities, laws for their connection, and means for studying them? That is, doesn't the old incommensurability problem rise again, this time much severely because it is replicated all the way down to an in-principle failure of communication even between subcultures?

This exacerbated incommensurability problem, it seems to me, forces us to re-assess the issue of locality.

Let's momentarily step back. It may well be that the most important change in our understanding of the history of science and technology was the recognition that accounts of technical work had to be local. The local configuration of laboratory work, pedagogical practices, and image-making techniques mattered. It mattered in seeing why forms of theory or experiment existed as they did; it mattered in grasping what counted as acceptable demonstration; it mattered in following how technologies and scientific work got used, altered, re-appropriated. The philosophical starting point of *Image and Logic* is that we ran into trouble when we tried to join two badly fitting notions. On the one side, we had all this fascinating, local work on practices—work emanating from a hybrid of historical, anthropological, and sociological inquiry. On the other side, we had a *global* concept of language: there was one kind of classical physics that one might, following Kuhn, think of as "Newtonian," and another, distinct and integral language, "Einsteinian." All the debates and struggles from the time of Kuhn up through quite a big piece of SSK tended to maintain this completely universal concept of the way language functioned. My interest in the development of ideas of scientific trading zones and exchange languages (scientific pidgins, jargons, creoles) aimed—and aims—to put local practices to work in a sea of localized scientific language. But back to the line of my work.

The third of my books on the subcultures of science was *Einstein's Clocks, Poincaré's Maps* (2003). Here, I aimed to take a particular statement ("simultaneity is nothing more than the same reading on clocks coordinated by the exchange of finite-speed light signals") and exploring the way that statement emerged from the triple intersection of philosophy, technology, and physics. By pursuing the assembly of theoretical and quite material practices, it is possible to provide an account of important aspects of science and technology that avoids the dead-end quarrels over whether science or technology are independent or dependent on some never-quite-defined "context." The abstract and the material enter together, from the start.

If one takes this localized view of theoretical, material and linguistic practices, once again problems of coordination become central: how do the subcultures of science and technology interact?

One can ask too: What kind of shared practices, language, symbol systems join them and how do these practical (not narrowly linguistic) inter-languages change over time? How do the most rarified of mathematical and physical concepts confront material systems altogether?

3. What, if any, practical and/or social-political obligations follow from studying technology from a philosophical perspective?

If technology is as much a knowledge system as science—and I think it is—then there is no sense in treating machines as empty vessels, objects outside interpretation. Here are some examples of ways I've been trying to grapple with the intersection of technology and politics. The first has to do with technical work, the others with technologies that shape matters more directly political.

Coordination without Homogeneity. We have to get away from two highly misleading models of how things get done in the collectivities that produce technical work. On one side, there is a tendency to think that homogenization is needed—that for biochemistry or nanotechnology to succeed, it requires the flattening of distinct disciplinary identities and cultures. On the other side, it is easy, too easy, to suppose that there must be a pyramid-like hierarchy in which one group or discipline plays a determinative role in all that follows. Neither the homogenizing view nor the hierarchical one helps much. We ought at least leave it as a contingent matter to discover what, in particular situations is shared as a trading zone between electrical, cryogenic, and mechanical engineers when they work together; or, for that matter, between nuclear weapon designers, computer designers, statisticians, and theoretical physicists. What objects or parts of objects are held in common? What rules of calculation and combination are exchangeable? What do theoretical physicists or even mathematicians draw from the concrete?

Technoprivacy and Secrecy. Privacy in modern legal reasoning dates in good measure from decisions made toward the end of the 19th century. At issue were, in no small measure, technological capabilities—the ability of a photographer to snap an image, or a penny newspaper to disperse accounts of "personal life" far and wide on short notice. Indeed, at every turn it seems that technology and privacy are bound up inextricably: aerial photography, infra-red surveillance, closed-circuit television cameras, data mining ... .the list grows wide and deep. One of my concerns has been to explore the ways in which shifting concepts of self cross with technologies to produce structures of far-reaching political implication. Privacy and government secrecy -which seem to vary inversely—are also closely bound. And here too, technology enters in fundamental ways. What epistemology must the secret-makers hold to act as they do, excising, blocking, forbidding the utterance of particular terms and propositions, images, materials, and machines?[1]

The Hydrogen Bomb. Back in the late 1940s and early 1950s, there was a singular moment in the history of weaponry. During World War II, the civilian scientists were, with but a handful of exceptions, so caught up in the total war, that there was little time devoted to moral and political reflection. After the mid-1950s, the military and atomic energy establishment had more or less weaned itself off the civilian sciences—so while debate could and was launched from the outside, the struggle of weapons designers themselves over what they were producing was somewhat moot. Between total war and the routinization of weapons production lay a narrow window—and it was in that window that the H-bomb debate of 1949-52 took place. From the trenches of the laboratories all the way to the top of the AEC and even to the inner circle of the President's advisors, the debate raged. Was the H-bomb the only response to the threat of expansionist Soviet Union—or was this weapon, by its very nature, and as the General Advisory Committee put it, necessarily a weapon of genocide that would spiral the human race into annihilation? Understanding the H-bomb is of course just to grasp one (particularly destructive) weapon. But confronting the philosophical problems it raises, about the responsibility of scientists, the inevitability of technologies, the transformations engendered in the practices of computers, simulations, are also part of its legacy.[2]

4. If the history of ideas were to be narrated in such a way as to emphasize technological issues, how would that narrative differ from traditional accounts?

In a certain way, my interest is not in the slow, trickle-down of ideas that eventuate in machines, but instead in the sudden confrontation of the ethereal and the material. So, instead of reading Richard Feynman as the quintessential "pure" theorist who spent the war banging bongo drums and cracking safes—take seriously the work he did on merchant calculators running routines over and over again to determine the flow of neutrons inside the critical assembly of plutonium or uranium-235. Following those forms of work leads to a different understanding of what he did after the war—in the seemingly unrelated domain of quantum electrodynamics. Or look at Paul Dirac, the theorist's theorist—and go back to his training in an "applied" Bristol technical school where he learned electrical engineering and a projective geometry aimed at producing a generation of highly-skilled workers. Turns out that Dirac, in some of his most productive "abstract" and purely algebraic quantum mechanical work, was, behind the scenes, working in a highly visual mode—one that emerged from his engineering and projective-geometrical youth.[3]

Better than the trickle-down account of thought becoming things would be, in my view, a constant confrontation of differing subcultures of work.

5. With respect to present and future inquiry, how can the most important philosophical problems concerning technology be identified and explored?

I am very reluctant to say this or that problem is the most important; more easily, I can say that I'm working to better understand the way certain technologies both depend on and reshape the self. For example, I'm trying to work out what the understanding of the self must be in place for Hermann Rorschach to have thought that ink blots could reveal in the inner structure of our conscious (and unconscious) perception. Then, once these tests become widespread—administered in the millions—they begin to reshape the self in new, and unexpected ways. If this is right, there are two moments to this "self-formation" process: first, the historical a priori that makes it possible for the technology even to be formulated; second, a contingent dynamic that unfolds as the technology gains traction. Another example: our nuclear wastelands presuppose a particular relation of us (humans) to nature—only with such a notion can one even formulate the idea of permanently "withdrawn" land that is forever off limits. But once such lands come into existence, once we drive ourselves out of nature, so to speak, (by radiological toxics and by legal stricture) it reshapes how we see ourselves in relation to the world. I've probably gone on long enough—for better or worse, this book-in-the-making is called "Building Crashing Thinking," and I hope it will be out in a year or so.

[1] On the hydrogen bomb, see e.g. Galison and Barton Bernstein, “In Any Light: Scientists and the Decision to Build the Superbomb,” Historical Studies in the Physical Sciences 19 (1989): 267-347; and Peter Galison and Pamela Hogan's film, “Ultimate Weapon: The H-bomb Dilemma.”

[2] Galison and Martha Minow, “Our Privacy, Ourselves in the Age of Technological Intrusions,” in Richard Ashby Wilson, ed., *Human Rights in the 'War on Terror,* Cambridge: Cambridge University Press, 2005, pp. 258-294; Galison, “Removing Knowledge,” *Critical Inquiry* 31 (2004): 229-43; Galison and Robb Moss, “Secrecy,” feature-length documentary film, in preparation for release in 2007.

[3] Galison, “Feynman's War: Modelling Weapons, Modelling Nature,” *Studies in History and Philosophy of Modern Physics* 29B (1998): 391-434; Galison, “The Suppressed Drawing: Paul Dirac's Hidden Geometry,” *Representations* 72 (2000): 145-66.

9

Allan Hanson

Professor, Department of Anthropology

University of Kansas
USA

---

1. Why were you initially drawn to philosophical issues concerning technology?

I began as a cultural anthropologist interested in Polynesia, first the small, remote island of in French Polynesia (where I did fieldwork for my dissertation and first book in the mid-60s) and then the New Zealand Maori. I was in a structuralist phase at the time, and traditional Maori culture thoroughly satisfied my thirst to detect general patterns beneath the surface forms of ritual, myth, folklore, art, and behavioral customs. By the middle 1980s, however, I found myself yearning to relate my research to the challenges of contemporary society. At the time my theoretical tastes were expanding to include poststructuralism. Foucault's *Discipline and Punish* was a seminal work for me. I was deeply impressed with its analysis of how material conditions and practices shape human lives. The section on the examination was particularly provocative, and Foucault's discussion of it as part of the disciplinary technology of power turned my interest to technology, broadly defined. In the margin next to Foucault's question, "Who will write the history of the examination?", my wife wrote "Allan will!" My major effort was *Testing Testing: Social Consequences of the Examined Life* (1993). Combining my structuralist and poststructuralist leanings, it sought underlying commonalities among diverse tests from lie detection and drug tests to vocational interest and intelligence tests, and analyzed them as disciplinary technologies of power in Foucault's sense.

One category of testing I did not include was medical tests. The original plan was to devote the next major project to that. That changed when the dramatic advances in new reproductive technologies in the early 90s and their massive potential to affect society led me to turn the study, and then I whittled it down to a focused analysis of the not-so-new reproductive technology of donor insemination. I was intrigued by the widespread speculation that, when it becomes feasible, people will use technology to improve their children's intelligence, athletic abilities, and other traits. Donor insemination has such eugenic potential, as is demonstrated by sperm banks advertising that their donors are Nobel Prize winners or otherwise gifted individuals. It struck me that an empirical test of the speculations regarding future use of reproductive technologies for eugenic purposes could be run by investigating actual criteria that have been used in selecting sperm donors. Moreover, donor insemination has important implications for feminism, because its growing use by single women and lesbian couples makes it easy for women to have children and form families without the active participation of men. I analyzed questionnaires from women who had used donor insemination, and interviewed many of them by telephone. The eugenic predictions were not well supported, and I found an array of attitudes toward the patriarchal family ranging from outright rejection to a desire to recapitulate it as closely as possible.

Probably the recent technological innovation that has impacted the lives of more people more profoundly than any other is the automation of information in the computer revolution. One cannot study technology very long without addressing it, and it became my next research focus. I will discuss some of what I learned later.

2. What does your work reveal about technology that other academics, citizens, or engineers typically fail to appreciate?

Although I have always been interested in the theoretical and philosophical aspects of whatever I study, I work more in the anthropology than the philosophy of technology. That means I am primarily interested in the cultural principles that inform technological practices, and their general consequences for individual and especially social life. In my work on testing, for example, I looked at a wide variety of tests and divided them into two categories. Authenticity tests are concerned to identify some state of being characteristic of an individual, a state that usually has moral or legal significance. Examples are lie detector tests and drug tests. Is this person telling the truth? Is he taking illicit drugs? Qualifying tests, on the other hand, measure a person's ability or inclination to perform certain activities. These include vocational interest tests, aptitude and intelligence tests, and all kinds of academic examinations (How much information can the student provide about 17th century England? How well can she solve algebraic problems?). The outstanding social consequence of testing is that it does not really assess pre-existing qualities. It manufactures them. Tests create what they purport to measure. As it is recognized and rewarded in social life, for example, intelligence is not so much a free-standing entity as an artifact of intelligence tests. This is the seldom recognized by massive social consequence of testing in contemporary society. Its capacity to form personal traits makes it a powerful technology of social discipline.

A major argument in my work on information technology is that it does not just enable us to do what we have done all along, only faster and more easily. It changes what we do and, most importantly, how we think. I am especially interested in how automated techniques for retrieving and assembling information are very good at indexing but very poor at classifying. They are extremely powerful in bringing together items that share a word or phrase, as in keyword searching (which amounts to constructing an index of the data set for the keyword query). But, unlike human intelligence, in the absence of such common terms, they cannot classify the data set into categories with similar content. This brings about a change in human thinking. Before automation most information was presented in already classified forms, such as encyclopedias, subject catalogues, and textbooks. For the most part, people accepted, worked within, and saw the world in terms of the pre-established classifications they had been taught and used. Information retrieved electronically does not come already classified, making it necessary for the human user to select from among the results of a search what is relevant and what is not ("garbage"). Sometimes the presence of items in automated search results that would not have been included in traditional classifying practices stimulates the human analyst to gain a new insight into the topic, or see it in a different way. This in my opinion is the most important unintended and little-recognized consequence of the automation of information: it presents the human mind with new challenges of interpretation and opens new possibilities for creative analysis.

Evidence for this shift is visible everywhere as fixed forms and institutions give way to greater flexibility of thought and action. In education and research the imperative to learn and work within the limits of one pre-packaged scholarly discipline is rivaled by interdisciplinary programs in universities and research projects that routinely cross disciplinary boundaries. Law and business firms now stress horizontal communication between departments as much as vertical communication within them, and they increasingly operate in terms of temporary task forces that bring together individuals with different areas of expertise for the purpose of dealing with particular problems and initiatives.

3. What, if any, practical and/or social-political obligations follow from studying technology from a philosophical perspective?

This question follows directly from the preceding one, because if the philosophical and cultural study of technology reveals unrecognized aspects with practical or social implications, the scholar should draw as much attention to them as possible. My work on the automation of information highlights at least two such implications. Both of them are pertinent to the culture wars.

One, by no means original with me, is that electronically mediated communication can fuel the culture wars. Television now has innumerable channels, many devoted to a single topic such as extreme sports or news with a particular slant. Internet blogs, list-servs, and chat rooms often vociferously promote one political or social point of view. All of these forms of narrowcasting make it possible to saturate oneself with information that reinforces one's biases and particular interests and blocks out other points of view. In this way electronically mediated communication can encourage the formation of isolated groups, many of which are hostile to each other, and that escalates the culture wars.

The other implication cuts in the opposite direction. I have mentioned already how automated information retrieval privileges indexing over classification, and how this challenges the human user to consider new possibilities and reach new insights. It follows that indexing tends to open minds, especially when compared with thinking confined within classification's pre-established categories. Polarization and the culture wars thrive on minds that are closed, that are neither interested in nor able to understand the Other. The best antidote is to understand and appreciate cultural differences. One important step toward that is the greater open-mindedness stimulated by automation's retrieval of information by means of indexing rather than classification, as already discussed. Especially important, this is not a matter of people deciding that they should be more open to the Other, as is the case, for example, with cultural relativism. It causes people to change their habits of using information. They don't think at all about whether they should be more open minded to people who are different from themselves, but nevertheless, that is a consequence of their changed habits. This is encouraging because different ways of thinking are more likely to become established when they result from changing habitual behavior than from direct efforts to convince people that they ought to change their minds.

If these two practical implications of information automation pull in opposite directions, which is stronger? People who are interested only in reinforcing what they already think will become more deeply entrenched with narrowcasting. They will become, if anything, more polarized in the culture wars. People who use electronic resources out of curiosity or in the pursuit of scholarly or business projects will find themselves becoming more open to new ideas and insights. This will position them to be more tolerant of cultural differences, and thus will be a force for calming the culture wars.

My current research is actually an attempt to evaluate the relative strength of these two positions. I am interested in the contention between absolutism and tolerance in contemporary American society. My initial focus is the law as a particularly interesting site for exploring this. It has the responsibility to umpire among disputing interests, but a debate exists within the law itself as to whether it should be entirely neutral or champion the Right Side. Those who hold the latter view are largely fundamentalist Christians who hold deeply committed positions on issues such as abortion and gay marriage. They promote their position through certain schools and legal foundations dedicated to training and practicing law with an evangelical and even theocratic mission. My goal is to learn how much of an impact this is having on our legal system and on society in general.

4. If the history of ideas were to be narrated in such a way as to emphasize technological issues, how would that narrative differ from traditional accounts?

The main difference is that the individual would not have enjoyed so prominent a place in explanations of action. Contrary to the widespread and long standing theory of methodological individualism, we are beginning to realize that the unit of action is not the human individual but something larger and less permanent, something that is often termed "cyborg," and which I call extended agency or the new superorganic. Any action should be understood as undertaken by a conglomeration that may include one or more human beings and other components such as plants, animals, machines, and other objects. The activity that I am participating in right now, for example, is done, at a minimum, by my embodied human intelligence (my thoughts expressed through my fingers as I type), a computer and a word processing program. Should I discuss it later with my wife, the pertinent agency will consist, again at a minimum, of our two embodied human intelligences (ideas expressed in audible language) and the printed-out text of what is now being written. Any activity is best described in terms of agencies with multiple components, that form, dissolve, and reform in different combinations to undertake other activities. Of the two activities just described, my intelligence participates in both (although differently embodied ... I'm typing in the one and talking in the other), but the other components (the computer, the print-out, my wife) participate in only one.

Because telephones, calculators, computers and other forms of artificial intelligence so obviously participate in what we do, technological issues have made it particularly clear that entities other than the individual human being must be taken into account in explanations of action. But this insight could, and probably should have been achieved much earlier in the history of ideas, because it is equally clear that activities in the preautomated era need also to be understood in terms of recombinant, extended agencies. This includes the interactions of multiple human beings, the participation of plants and animals, and the use of tools going back to the Paleolithic. Had this been grasped earlier, the history of ideas would have been different, and human hubris would probably not have gotten so out of hand.

5. *With respect to present and future inquiry, how can the most important philosophical problems concerning technology be identified and explored?*

Certainly many issues will command the attention of future researchers, among them the social consequences of nanotechnology, biotechnology, robotics, and the implications of computer-mediated communication for human relationships and community.

I will be particularly interested in research on how further developments in reproductive technologies will impact heterosexual and homosexual relationships, marriage, family structure, and the possibility of a new eugenics.

Another critical issue, and one that I am currently working on, is the implication of cyborg or extended agency theory for responsibility. With methodological individualism, allocation of responsibility was simple, probably overly so: if human individuals are the authors of action, then they are responsible for it. But if we are now to recognize that the agencies that do things extend beyond people to include nonhuman components, where does responsibility lie? With the agency as a whole? With one or a few parts of it? Who or what should be blamed or praised? What form should rewards and punishments take? An example all too familiar to teachers is students' current excuse of preference for why term papers are late: "my hard drive crashed." Often this does not get them very far, but in fact it is more plausible than the earlier favorite—the dog ate it—because today hard drives are in fact essential to the production of term papers while dogs have never been.

This is something, as I say, that I am currently working on. The job is far from complete, but I think the most fruitful way to approach it is to recognize that we already apply responsibility-linked concepts such as "fault," and "blame" and "approval" to inanimate objects, machines, animals, children and corporations as well as to individual adults. Analysis of those usages reveals that responsibility and the related sanctions represent a family of ideas with variations that depend on the type of being in question. My hunch is that, considered in this light, extended agencies will not require a radical revision of the notion of responsibility that we presently use.

Selected Bibliography

The Trouble with Culture: How Computers Calm the Culture Wars. Albany: State University of New York Press, forthcoming.

"From Classification to Indexing: How Automation Transforms the Way we Think". *Social Epistemology 18*:333-356, 2004.

"The New Superorganic". *Current Anthropology 45*:467-482, 2004.

"From Key Numbers to Keywords: How Automation Has Transformed the Law". *Law Library Journal 94*:563-600, 2002.

"Donor Insemination: Eugenic and Feminist Implications". *Medical Anthropology Quarterly 15(3)*:1-26, 2001.

"How Tests Create What They Are Intended to Measure". In *Assessment: Social Practice and Social Product*, Ann Filer, ed., pp. 67-81. London: Routledge/Falmer: 2000.

"Testing, The Bell Curve, and the Social Construction of Intelligence". *Tikkun 10(1)*:22-27, January/February 1995.

"The Invention of Intelligence". *Education Week*, September 15, 1993, pp. 40, 32.

Testing Testing: Social Consequences of the Examined Life. Berkeley: University of California Press, 1993.

"Le dépistage des drogues: contrôle des drogues ou des esprits?" *Psychotropes 7(3)*:71-87, 1992.

"What Employees Say About Drug Tests". *Personnel*, July 1990, pp. 32-36.

"Performances mentales et définition de la personne". *Actions et Recherches Sociales, No. 2*, juin 1990, pp. 31-36. (Issue titled Identités de papier.)

"Some Social Implications of Drug Testing". *Kansas Law Review 36*:899-917, 1988.

"The Lie Spy's Flaw". *Explore*, Winter 1988, p. 16.

Counterpoint in Maori Culture (with Louise Hanson). London: Routledge & Kegan Paul, 1983.

Rapan Lifeways: Society and History on a Polynesian Island. Boston: Little, Brown, 1970.

10

Donna J. Haraway

Professor, History of Consciousness Department

University of California at Santa Cruz
USA

---

1. Why were you initially drawn to philosophical issues concerning technology?

If by "philosophical issues concerning technology" one means questions about what ways of living and dying, what ways of making and experiencing meanings, and what sorts of mind/bodies who gets to have and under what conditions, then I was initially drawn to these inquiries shortly after leaving the haploid state and just before uterine implantation. This kind of answer signals the sort of philosophy of technology that I inhabit and that inhabits me; i.e., a historically situated, bioscientific, 'western' approach attuned to always relational organic bodies as technologies. My philosophical mode of attention is also obviously shaped by the likes of the great cartoonist, Gary Larson, whose Farside animal jokes pepper the doors of biologists and poke holes in the brains of the literal minded. Locating my origin story on the hiatus between haploidy and implantation also signals my indebtedness to the great theorists of woman the gatherer and other genealogies foregrounding motherlines. Faced by the relentless accounts of the Fathers of technology and philosophy, I must be forgiven a little old-fashioned, unmodified, woman-centered feminism, even when I know 'woman' to be one of the most problematic technical fabrications in the western lexicon.

Let's get back to those dramatic originary moments for the knowing subject suspended between ovoid haploidy and uterine attachment. Primed by reading Heidegger and Marx in college, my inquiries about the philosophy of technology suffered a rude torque in graduate school at Yale in cell and developmental biology when I tried to tell my peers that the cell and the embryo were, in some strong sense I did not know how to articulate, "invented" rather than "discovered." They thought I meant that the plainly visible (granting help from snazzy microscopes and a few other toys) cells and embryos were "made up" and that I was either joking or crazy. I thought I was trying to figure out how come—and how—biologists studied really real organic entities, ones I was in love with, which looked and acted a lot like heat engines, electronic circuits, systems ordered by the hierarchical division of labor parsing reproduction and production, stratified executive powers, communications devices, racial and national formations, war and market strategies, and game-theoretic algorithms. I was also riveted by biologists who studied developing embryos in the same way I had learned to read poetry in college, as manifestations of dynamic organic form. I was also horrified by some mentally challenged biologists who appeared to study organisms as bean bags of genes. Later, I learned in detail how scientifically educated people who consider themselves ethical and rational turn other living animals into nothing but well managed factories on the hoof or claw, literally and tropically. And through it all, I only fell deeper in love with the many biological sciences as well as with their plucky organisms.

In short, I was in real trouble as a biologist and utterly unprepared for anything else. Reduced to diploidy by spermatic enhancement and tunneling down the Fallopian tube toward the bloody comfort of gestation, I clutched to my root insight: culturally, politically, and historically located metaphors, models, and other tropes had a great deal in common with juicy material things in the world, and especially with organisms. Built tropes became my lifeline; and I knew for certain that all the builders are not and never have been human. It is no wonder that my PhD in biology was granted for a thesis that came out as a book called *Crystals, Fabrics, and Fields: Metaphors of Organicism in 20th- Century Developmental Biology* (New Haven: Yale University Press, 1976; reprinted as *Crystals, Fabrics, and Fields: Metaphors that Shape Embryos* (Berkeley: North Atlantic Books, 2004).

2. What does your work reveal about technology that other academics, citizens, or engineers typically fail to appreciate?

An ugly duckling called a cyborg issued from a very U.S. American, multi-lobed womb that included a scientific education even for devout Catholic females financed by Cold War prosperity; the ethereal electronic battlefield coupled to the muddy body counts of Viet Nam; anti-nuclear movements on Native American land for Mother's Day; scientists organizing for science for the people; civil rights movements, Angela Davis, and the Black Panther Party; gay liberation and Adrienne Rich's lesbian continuum; a blend of Rachel Carson, Ursula LeGuin, and the techno-ecosystems of Eugene and Howard Odum; new-left Marxism welded to liberation theology; Samuel R. Delany's science fiction read with wanton ironic literalness; and a hopelessly flawed, highly creative, middle-class, all-too-white feminism in the company of many other feminisms. Given to provoking trouble for kin and kind, this ungainly girrrl (sic) offspring called cyborg was disloyal to gender and wary of organicism; and s/he was not initially recognized as one of us in my matrilines. Worse, she was all-too-legitimate for some wired fathers and would-be brothers who—editing out both the angry and the joyous feminist—crowed about a blissed-out female technochick in the flock.

But for me, besides relentless insistence on the material-semiotic nature of scientific and technological beings of human and non-human kinds, the cyborg is about the fact that contradictory things are simultaneously true all the time. No philosopher really likes contradiction as the endpoint of inquiry; that's why I am a bad philosopher. Without even dialectical resolution, the sheer messiness of life—and of technology—seems our best hope for breaking the hold of the established disorder. The world is not finished; and reconfigured knowledges and technologies must be at the center of freedom projects, including feminism. Serious anti-racist feminists, in particular, cannot afford to be "for or against" science and technology; ways of life are at stake inside science and technology. One cannot know in advance what something is, not even or maybe especially if that something comes from the belly of the monster. I think what my readers most mis-understand about technology is that it inescapably and simultaneously fuels both joy and rage, analytical argument and implicit knowledge, refusal and reinvention, tripping and skill. With no warrant to be for or against the techno-organic weld that is material-semiotic reality for countless human and non-human beings on this earth, we can then get it that the world is in play and at stake. Response is non-optional.

Furthermore, technology is not a set of tools made and used by power- and wealth-stratified people in the history of a species 'progressively' dominating (or destroying) its planet. Rather, technology is a relational matrix of humans and non-humans that shapes and is shaped by skill, dream, need, power, community; there is no outside and there is no narrative structure with origins followed by good or bad endings. Technology is about what developmental biologists call 'reciprocal inductions'. These infoldings and inductions shape who and what lives and dies and how.

3. What, if any, practical and/or social-political obligations follow from studying technology from a philosophical perspective?

Any decent philosophy of technology should elucidate and enrich "docking sites"; i.e., possible points of entry and attachment for an indeterminate crowd of actors committed to techno-organic flourishing on this planet. A philosophy of technology should make its practitioners and constituencies more worldly, more engaged in building, unbuilding, and rebuilding the material-semiotic flesh of mortal lives in all of the physical-chemical and political-cultural media necessary to flourishing. If technology is about reciprocal inductions and infoldings of human and nonhuman flesh for building ways of living and dying, then such shapings must be examined for crucial practical implications—like whose skills are valued and improved and whose shipped to the landfills of history? What would it take to build connections across a chasm separating different communities of practice which 'philosophically' need each other? How can human and 'more-than-human' beings, landscapes & seascapes, built environments, and tools and artifacts be the subject of respect, heard as respecere or looking back, looking again, and holding in esteem and regard? Can respect be more of a guide to action than critique? What would the timeliness of innovations in a field of engineering look like if the whole sweep of processes including imagining, designing, building, assimilating, using, wearing out, discarding, and recycling were figured into the account in grainy, fleshly, and financial detail? What would bioethics look like if the questions were posed less in regulatory and restrictive terms, and more in terms of expanded 'democratic' discussions about flourishing in necessarily contradictory conditions? What would enable such conversations—financially, intellectually, emotionally, culturally, politically? If philosophers of technology know something these days about communities of practice and how things (in Latour's sense of things) get co-constituted—and we do know quite a bit—then the practice of philosophy itself is part of the mix and actually can't escape its own part in making some worlds more possible than others.

So, the issue is not so much whether there are practical or socio-political obligations flowing from the philosophy of technology as how to face up to the fact that we are always already engaged and had better get down and dirty in the mess of things, with our very best expert and personal selves on the line, and not imagine there is any other choice. It's not about obligation so much as honesty. Who and what lives and dies inside a relational process called technology is an eminently practical question that requires research, emotional and cognitive vulnerability, the discipline of making mistakes interesting, staying at risk deep inside oneself and one's communities so that more than what is already in one's ken can have a better chance to flourish in the built world.

4. If the history of ideas were to be narrated in such a way as to emphasize technological issues, how would that narrative differ from traditional accounts?

"Ideas" are themselves technologies for pursuing inquiries. It's not just that ideas are embedded in practices; they are technical practices of situated kinds. That said, there is another way to approach this question. Several years ago I took a freshman course on American History offered at night at our local community college in Healdsburg, CA, in order to add to the enrollment figures so that the instructor, my husband, Rusten Hogness, could give me an "F" and thus have the freedom to give better grades to the real students, since the history department insisted on grading to a strict curve. Among other pursuits, Rusten is a software engineer who then was working at a little Hewlett-Packard branch office with fellow engineer friends. They all took the course for failing grades too, so that Rusten and his students could forget the curve and concentrate on learning. Without giving away our identities or purposes to the other students, who were of varying ages and experiences, all of us rogue enrollees actually worked pretty hard and joined in discussions all the time. Rusten taught the whole course through the history of technology, focusing on things like the shoe lasts, guns, surgeries, and potted meat of the Civil War; the railroads and mines of the Rocky Mountain west; the calorimeters of food science in land grant colleges and their relation to labor struggles; and P.T. Barnum's populist testing of the mental acumen of visitors to his displays (were they a hoax? were they real? I seem to recall that to be a famous philosophical query). Throughout the class, a wide ranging set of questions in philosophy, politics, and cultural history came together to think better about possible shapes of science and technology. The idea that technology is relational practice that shapes living and dying was not an abstraction, but a vivid presence. The history of a nation, as well as the history of ideas, had the shape of technology. Old and important books like *Mechanization Takes Command* and *Technics and Civilization* helped us through the course's conventional required textbook. The real students, as well as the faux failures, loved the course and knew a great deal more about "the history of ideas," including things like information and thermodynamics, as well as work, land rights, and justice, at the end of the term than at the beginning.

Rusten loves to teach, and he is fiercely committed to democratic scientific and technical competence and literacy. He has always taught with as much of a hands-on approach as possible and with a bright eye on the history of popular science and struggles for a more democratic society. We met in the 1970s in the History of Science Department at Johns Hopkins where he was a graduate student studying 19th-century French and American popular science, among other things. He was also teaching the natural sciences and mathematics, as well as history and social studies, at the Baltimore Experimental High School. There, he constantly had his students hanging out in labs, hospitals, factories, and technology museums; and he taught politics, history, science, and technology as an integral part of Baltimore's story as an industrial port city with a fraught racial, sexual, and class history. He turned our kitchen into a chemistry lab, literally, and got the students thinking about industrial chemistry as well as the science and technology of cooking as a way to nurture both the pleasure of the science and a better sense of how divisions of labor and status work in science and technology.

Years before, Rusten, a war resister and pacifist in the Viet Nam era, had done two years of alternative service in the southern Philippines, teaching mathematics and philosophy in a little fisheries college to students who were mostly dead a few years later from the repression of revolutionary movements by the U.S.-supported regime in Manila. Questions about technologies of globalization and of "anti-terrorism" are indelibly written onto his optic tectum and in intimate contact with whatever signals are working their way through the cerebrum.

Rusten's paternal grandfather had headed the physical chemistry division of the Manhattan Project and then participated in civilian scientists' struggles over the control of nuclear science and technology after the war. Perhaps as a result, most of Rusten's siblings and cousins are directly engaged in their jobs in "the history of ideas from a technological perspective" and vice versa. I tell this family story to foreground the knot of public and intimate worlds tying together what we call technology and what we might mean by philosophical perspectives. I am not sure if this way of approaching the question is traditional or not; it depends on what tradition one focuses on. But I am sure that I learned more U.S. history and more history of philosophy, as well as history of technology, in the one course in my life that I failed than in a great pile of those others marked with "A's."

5. With respect to present and future inquiry, how can the most important philosophical problems concerning technology be identified and explored?

There is nothing very mysterious about how to identify important philosophical problems concerning technology. They press on us in such a way that avoiding them seems the more puzzling problem. Some questions tied to technology are more urgent than others, questions that seem to lead back to critique in the 'negative' sense, but really should provoke us to reimagining what might still be possible. Matters that stay pretty high on the list include the technologies of war, environmental degradation, technical apparatuses for trashing vast numbers of people and other organisms, making and shaping children, sorting people into categories like vermin and CEOs, and on and on. Some questions are also full of promise for adding to the stock of joy, sensuality, and curiosity for those who have not had a great deal of permission to nurture such capacities; and these questions are also urgent if perhaps harder to identify. In any case, both the urgent troubles and the urgent openings must be addressed from a deeply felt thinking and acting from inside always already ongoing practices of collectively building and destroying worlds, and not in the abstract.

Books

When Species Meet (Minneapolis: University of Minnesota Press, forthcoming 2007).

The Companion Species Manifesto: Dogs, People, and Significant Otherness (Chicago: Prickly Paradigm Press, 2003).

The Haraway Reader (New York: Routledge, December, 2003).

Modest Witness@Second Millenium. FemaleMan Meets OncoMouse: Feminism and Technoscience (New York and London: Routledge, 1997).

Simians, Cyborgs, and Women: The Reinvention of Nature (London: Free Association Books and New York: Routledge, 1991).

Primate Visions: Gender, Race, and Nature in the World of Modern Science (New York and London: Routledge, 1989; London: Verso, 1992).

Crystals, Fabrics, and Fields: Metaphors of Organicism in 20th-Century Developmental Biology (New Haven: Yale University Press, 1976; reprinted as *Crystals, Fabrics, and Fields: Metaphors that Shape Embryos* (Berkeley: North Atlantic Books, 2004).

Selected Articles

"Animal Sociology and a Natural Economy of the Body Politic, Part I, A Political Physiology of Dominance," *Signs 4 (1978):* 21-36.

"Animal Sociology, Part II, The Past Is the Contested Zone: Human Nature and Theories of Production and Reproduction in Primate Behavior Studies," *Signs 4 (1978):* 37-60.

"The Biological Enterprise: Sex, Mind, and Profit from Human Engineering to Sociobiology," *Radical History Review, no. 20*, (spring/summer, 1979): 206-37.

"The High Cost of Information in Post World War II Evolutionary Biology: Ergonomics, Semiotics, and the Sociobiology of Communications Systems," *Philosophical Forum XIII, nos. 2-3 (1981-82)*: 244-78.

"Signs of Dominance: From a Physiology to a Cybernetics of Primate Society, C.R. Carpenter, 1930-70," *Studies in History of Biology 6 (1983)*: 129-219.

"Manifesto for Cyborgs: Science, Technology, and Socialist Feminism in the 1980s," *Socialist Review, no. 80 (1985)*: 65-108.

"Situated Knowledges: The Science Question in Feminism as a Site of Discourse on the Privilege of Partial Perspective," *Feminist Studies 14, no.3 (1988):* 575-99.

"The Biopolitics of Postmodern Bodies: Determinations of Self in Immune System Discourse," *Differences: A Journal of Feminist Cultural Studies 1, no. 1 (1989):* 3-43.

"The Promises of Monsters: Reproductive Politics for Inappropriate/d Others," Larry Grossberg, Cary Nelson, and Paula Treichler, eds., *Cultural Studies* (New York: Routledge, 1992), pp. 295-337.

"A Game of Cat's Cradle: Science Studies, Feminist Theory, Cultural Studies," *Configurations: A Journal of Literature and Science 1 (1994)*: 59-71.

"Cloning Mutts, Saving Tigers: Ethical Emergents in Technocultural Dog Worlds," in *Remaking Life and Death: towards an anthropology of the biosciences*, Sarah Franklin, and Margaret Lock, eds. (Santa Fe, NM: School of American Research Press, 2003), pp. 293-327.

"Cyborgs to Companion Species: Reconfiguring Kinship in Technoscience," in *Chasing Technoscience: Matrix of Materiality*, Donald Ihde and Evan Selinger, eds. (Bloomington: Indiana University Press, 2003), pp. 58-82.

"Crittercam: Compounding Eyes in NatureCultures," in *Postphenomenology: A Critical Companion to Ihde*, ed. Evan Selinger (Albany, NY: SUNY Press, 2006), pp. 175-88.

11

N. Katherine Hayles

John Charles Hillis Professor of Literature
Distinguished Professor

Departments of English and Design Media Arts
University of California, Los Angeles
USA

---

1. Why were you initially drawn to philosophical issues concerning technology?

I began my professional life as a scientist and only then switched to the humanities. The juxtaposition of the two fields encouraged me to think in specific and material terms about technology, with questions such as these: How is it made? How does it work? What interfaces and affordances does it provide? My humanities education encouraged me to focus on a very different set of questions: What are the effects of building this system rather than another? What are the cultural and social implications? What does this technology mean for human experience and subjectivity? The intersection of these two orientations is where my work is situated, a location akin in many ways to philosophical inquiry. I do not, however, consider myself a philosopher but rather a denizen of trading zones where the conjunction of the technological and humanistic becomes not simply desirable but absolutely necessary to approach the full complexity of the situation.

2. What does your work reveal about technology that other academics, citizens, or engineers typically fail to appreciate?

Much of what I write is more or less obvious from one perspective or another. My contribution, as I see it, is to bring diverse perspectives together within the same work so that, for example, engineers find familiar insights but also new information that to humanists would be familiar, while humanists experience the inverse. The genuinely new, insofar as it inhabits my work, comes from the friction generated by juxtaposing diverse perspectives. For example, in *How We Became Posthuman*, my archival research into the Macy conferences put this familiar material in a context where it could be understood in larger terms as a confrontation of liberal humanist subjectivity with a theoretical framework that subverted its presuppositions. Without the specificity of the technological artifacts associated with cybernetics, that confrontation could not have become explicit; without the cultural analysis of liberal subjectivity, the full meaning of what those artifacts signified in their cultural context could not have been articulated. It was the friction between the artifacts and the ideology, and beyond that, the assumptions of a technological elite with broader cultural currents, that generated the seminal insights of my study.

3. What, if any, practical and/or social-political obligations follow from studying technology from a philosophical perspective?

My experience in science gave me a first-hand appreciation for the pressures under which scientists typically work, especially in cutting-edge areas where the competition for funds, human and technological resources, and prestige is intense. In such environments, it is not surprising that researchers do not have a lot of time to ponder the effects of their discoveries as they spin out into the culture. While there is a long tradition of scientists who do think about these matters—we may recall, for example, Albert Einstein's attempts to warn leading politicians about the implications of nuclear weapons—such interventions are probably more the exception than the rule. Moreover, fully understanding the cultural and social implications of a nascent technology is especially difficult, because it may develop in completely unforeseen directions. Given these constraints, humanistic inquiry should take an active role in thinking along with scientists and engineers about difficult questions concerning the social and cultural implications of developing technologies in both national and global contexts. A robust dialogue between humanists on the one hand, and scientists and engineers on the other, can potentially lead to deeper insights on both sides and, in the optimal case, influence the directions in which the technology develops. A good example is the collaboration between AIDS activists and medical researchers. The activists paid their dues by learning enough of the science to communicate effectively with the researchers, and the results included salutary changes in research protocols, for instance by eliminating the use of placebos so that no one was denied the appropriate medical intervention. The relationship between those who create and develop the technologies and those who critique it is not always easy or comfortable, but in my view it is absolutely necessary to ensure that we move in life-affirming and life-enhancing directions.

4. If the history of ideas were to be narrated in such a way as to emphasize technological issues, how would that narrative differ from traditional accounts?

The very phrase "history of ideas" carries a powerful implication that the important part of any history lies in concepts. Among the many problems with this assumption is the implication that ideas can be separated from their material and experiential contexts, as if they were entities in themselves. In practice, however, concepts always develop in embodied relationships with materiality, including the physical arrangements of laboratory space, the potentialities and intractabilities of instruments, and the resistances and enablings that the stuff of the universe offers when humans attempt to manipulate it. Andy Pickering has written eloquently about the interaction between concept and materiality in *The Mangle of Practice*, showing through case studies how feedback loops between ideas and materialities are central to the development of scientific theories.

Narratives written with these interactions in mind look very different from traditional histories of ideas, focusing not just on the development of concepts but on artifacts, laboratory spaces, networks between researchers, instrumentations, cultures of research sites including hierarchies, protocols, and lines of communication—in general, the practices by which a research community constitutes itself as such. So rich and extensive have been these kinds of approaches in the last thirty years that it now sounds almost quaint to refer to "the history of ideas," as if we had entered a time warp and been transported back to the era of fins on automobiles and suburban housewives at coffee klatches. The world is more than ideas and always has been. To understand the place of ideas in the world, context and materiality are crucial.

5. With respect to present and future inquiry, how can the most important philosophical problems concerning technology be identified and explored?

As I indicated above, I do not see philosophical problems as a separate realm from material embeddedness and cultural context. A full appreciation for the complexities of a given situation requires attention to all these aspects and more. Given that an individual's effort will be limited by her particular experience and expertise, collaboration appears especially attractive. I would love to see more collaborations between humanists and scientists and engineers. Each community can be enriched by the insights of the other. What the scientist thinks might be the important problems will likely look different from a humanistic perspective, and what the humanist assumes to be the case is often challenged and transformed by an intimate acquaintance with the work of scientists and engineers. To make such collaborations fruitful, mutual respect is necessary—respect that nevertheless thrives on tough questions and close encounters in which the friction of different viewpoints can be converted into mutually illuminating insights. I am thinking especially of those moments when each side suddenly understands the context in which the perspective of the other makes perfect sense. One need not be persuaded by such moments to change one's own viewpoint, but in my experience, these moments lead to enhanced understanding and open the door to deeper and more meaningful conversations.

Speaking especially of literary studies and other disciplines whose methodologies are primarily discursive, I have observed among my colleagues the inclination to think that words trump material reality. Indeed, I can recognize the inclination so clearly because I share in it myself. For a researcher steeped by contrast in material culture, things speak much louder than words. We could argue endlessly about the power of discourse to shape material reality, while those on the other side of the Grand Canyon can yell the opposite sentiment across the chasm. Things only start to get interesting, however, when the echoes mingle and interact with each other, setting up more complex patterns than either side could generate on its own. Hearing those interactions, we might be moved to adjourn to another setting where we could talk without yelling and hear one another more clearly.

I have been attempting to participate in and perhaps even initiate such intercultural dialogue for some time. In my recent work I have explored the conjunction of materiality and discourse in the context of electronic textuality, where language and code, digital and analogue signification, and human and machine cognition dynamically and recursively interact with one another. On the one hand, the computer can be seen as an abstract logic machine; on the other, it is a material object embedded in networks through which circulate messages, resources, and people. Seeing it as both a material technology and a symbolic processor instantiates the kind of perspective that in my view can move the conversation forward.

Citations

Clark, Andy. *Natural-Born Cyborgs: Minds, Technologies, and the Future of Human Intelligence.* New York: Oxford University Press, 2004.

Haraway, Donna. *Modest Witness@Second Millennium. Female-Man Meets OncoMouse: Feminism and Technoscience.* New York: Routledge, 1997.

Hayles, N. Katherine. *How We Became Posthuman: Virtual Bodies in Cybernetics, Literature, and Informatics.* Chicago: University of Chicago Press, 1998.

—*My Mother Was a Computer: Digital Subjects and Literary Texts.* Chicago: University of Chicago Press, 2005.

—*Writing Machines.* Cambridge: MIT Press, 2002.

Hansen, Mark B. N. *New Philosophy for New Media.* Cambridge: MIT Press, 2006.

Hutchins, Edwin. *Cognition in the Wild.* Cambridge: MIT Press, 1996.

Kirschenbaum, Matthew. *Mechanisms: New Media and Forensic Textuality.* Cambridge: MIT Press, 2006.

Latour, Bruno. *We Have Never Been Modern.* Cambridge: Harvard University Press, 2006.

Pickering, Andrew. *The Mangle of Practice: Time, Agency, and Science.* Chicago: University of Chicago Press, 1995.

12

Don Ihde

Distinguished Professor of Philosophy

Department of Philosophy, Stony Brook University, New York
USA

---

1. Why were you initially drawn to philosophical issues concerning technology?

In my first post-Ph.D. appointment at Southern Illinois University (1964), I became part of an interdisciplinary honors program. Its theme was 'the leisure society' under the notion that technologies would free up vast amounts of leisure time and the implied utopian hope was that this would lead to a 'New Greece.' That did not seem to match my own experience or what could be noted in then dominant social trends—so I began to do a "phenomenology of work." From this emerged the realization that much work, even academic work, engaged technologies and other material based processes. For example, how do we compose our written results; how are these disseminated; what is the material culture in which we are engaged? Some years later, after initiating the first philosophy of technology course at Stony Brook (1971), I would ask students to keep a diary of just how many human-artifact uses they engaged in during a single day. Results were astonishing for everyone. Back at Southern Illinois at that earlier time three books I used were suggestive: Herbert Marcuse's *One Dimensional Man* was almost cultishly popular amongst undergraduates, and I used it in several classes. It was clearly dystopian in character and blamed technology for much woe in the then contemporary world. Hannah Arendt's *The Human Condition* was also influential, but only indirectly. What I got from her was more about the important role of slavery in classical Greek society. And, finally, I had begun to teach Martin Heidegger's *Being and Time* which did have, in his famous tool analysis, a deeper analysis of the role of technologies in human history and society. Interestingly, there was a sense in which Marcuse and Arendt showed something quite contradictory about technologies in societies: Marcuse, in effect, blamed modern 'slavery' upon technology; but Greek slavery existed in what was clearly a low-tech and even technology-poor society. Something had to be wrong here. Those were the 60's, and by the 70's I had become seriously interested in philosophy of technology.

2. What does your work reveal about technology that other academics, citizens, or engineers typically fail to appreciate?

Looking back now over four decades of reflection upon human-technology interrelationships, I think one slowly and gradually can develop a sensitivity to the subtle, often overlooked but clearly important roles that the many technologies we engage and experience display. My own work, phenomenological in character, shows how this interaction is such that just as we invent and shape our technologies, the reverse is also the case and we are shaped by our technologies. And, I hold, this is the case for the simplest technologies as well as for larger more complex ones. To use a dip ink quill to write constrains how fast and how one can shape one's writing and I have to accommodate myself to the constraints and possibilities of the quill pen. Similarly, with a brand new, high tech car, one has to learn how to embody its outlines, extent, and speed capacities so as to park and pass properly. Early on, I began to be fascinated particularly by instrumentation—and while much of my career has focused upon the role of science's instruments, other instrumentation such as in media and in music also played roles. Instruments are particularly relevant for human perception and embodiment and it has often been this aspect of human-technology relations which I have worked upon. Science interpretation, intellectually, sometimes likes to characterize itself as disembodied or not concerned with embodiment, but when one looks at its practices in experimentation and its uses of instruments, very strong senses of embodiment are implied. For example, imaging technologies even if one is imaging phenomena which lie beyond human sensory capacities, must translate these back into imagery which can be seen and interpreted, thus indirectly implying human embodiment. Musical technologies, instruments, are often interpreted in a different way. Here bodily skill is of clear foreground importance and in the history of music there is often a resistance to the introduction of new instruments, precisely because of worries about bodily deskilling possibilities. Similarly, the way technologies become culturally embedded differs from time to time and from place to place, sensitivity to such material cultural patterns is also called for. What philosophers do is to look for patterns and then reflect upon the implications of such patterns within human individual and social experience. I have used the notion of a technological trajectory in many of my case studies or historico-empirical investigations. A simple example is in optics: one trajectory of optics is the phenomenon of magnification. Once inventors find out how to magnify things, a developmental trajectory is suggested: more magnification shows more interesting things or aspects of things. Thus the history of telescopes follows just such a trajectory, from Lippershey's original 3 power compound telescope, to Galileo's ultimately 30 powered one, later again improved by Newton's new reflection telescope which again increased powers ... and on into the contemporary ones which discern and depict the cracks on the icey surface of Europa, one of Jupiter's moons. The opposite magnification trajectory into the microscopic world also was followed, to the point today that even atoms are imaged by electron microscopy. But if these are technological trajectories, they are also more: I have argued that science often follows its instrumental trajectories. Microphysics and genetic biology, frontier science areas, are only possible with the instrumentation which allows the science to see and manipulate the objects of those sciences.

3. What, if any, practical and/or social-political obligations follow from studying technology from a philosophical perspective?

One important realization that emerges from the philosophical reflections upon technologies, and I think could be said to be a matter of consensus among most philosophers of technology today, is that all technologies are non-neutral. Indeed, for any technology to be interesting, it has to change something—if it left everything exactly as it was, it would not be interesting. This means, minimally, that any invention will have some social and political consequence. But, here one runs into a major problem: on the one side, such consequences are complex, and I have argued, multistable. Thus there will remain inherent in technological innovations, a recalcitrant indetermination and unpredictability. On the other side, precisely because technologies-in-use are non-neutral and change conditions, possibilities and the like, we should be responsible for reflecting upon such changes. I have always argued against taking either extreme utopian, or dystopian stances, but instead have argued that the ambiguous and multistable situation which technologies pose calls for careful and critical reflection, and reflection very much focused upon the empirical or concrete situations posed by specific technologies. Many of these effects are also doubled and tripled in consequences—Peter Paul Verbeek argues that sonography which images the human fetus leads to a different set of understandings about the roles of pregnant mothers and their developing fetuses and thus changes both social understandings and medical practices. But at the same time, there are beginning to be signs that repeated sonograms may also not be neutral with respect to the even minimal ultrasound waves which penetrate the fetus. There is an emergent question about whether or not the development of the central nervous system may be affected by ultrasound waves if these are applied frequently. In the same medical context, can anyone doubt that the laproscopic surgery, aided by fiber optics and microtools, is a vast improvement upon the older much more invasive and radical surgeries which preceded microsurgery?

4. If the history of ideas were to be narrated in such a way as to emphasize technological issues, how would that narrative differ from traditional accounts?

Were all histories to be narrated with a sensitivity to material culture and technological embeddedness, I have no doubt much would change in our understanding. In recent years I have been trying to develop what I call a material hermeneutics. Traditionally, hermeneutics has been usually thought of as a style of critical interpretation focused largely upon linguistic related phenomena: texts, language, speech, in short the linguistically meaningful phenomena which concern most humanities and social science disciplines. A material hermeneutics would, in contrast, focus upon materiality and metaphorically "letting the things speak". Put most simply, this would call for humanities and social science practitioners to utilize the same material investigative technologies employed in the natural sciences. One of my favorite examples relates to the 1991 discovery of Otzi, the 5300 year old "Iceman" found in the Italian Alps. Indeed, much of the knowledge about

Otzi continues to fall into the knowledge familiar to archeologists and anthropologists as evidenced by the objects of his material culture found with him. He had a partially completed archery set (bows and arrows), an axe, which turned out to be copper and older than any known use of that metal previously known in Europe, a medicine and a fire making kit, etc. Yet to date him and his artifacts, Carbon 14 techniques had to be used with the result that he was found to be much older than initially thought to be (5300 BP). Then, with techniques only developed in the 20th century, his death could be timed via paleaopathology—his stomach contents included pollen of the hop hornbeam tree which sheds pollen only in late spring, charcoal fragments and eikhorn wheat or bread remains, and with DNA, red deer and mountain goat particles were found. In short, the material hermeneutics made possible by instruments such as mass spectroscopy to electron microscopy, gives us a richer tale concerning Otzi than of most many millennia later humans. Were this kind of analysis to be part of our intellectual histories, our histories would certainly be richer and more complete.

The same would be the case on historical-social levels. Although Karl Marx hinted in the 19th century that different modes of production yielded differently shaped societies (in what today seems too simplistically deterministic a form), more detailed attention to technology-society relations would be important. If the theorists who think ours is an information or a knowledge society which is different from older forms of society, specific studies are called for to take account of just how this is so. It does seem clear that the older notions that societies must progress through the various stages of technological development in some kind of order, is no longer clearly the case. For example, software development, even hacking and scamming is clearly not restricted to 'first world' countries in the contemporary world; it takes place in many developing countries as well. Instead, miniaturization, distribution, and regional development makes 'leapfrog' adaptations possible virtually everywhere. Similarly, the complex networks which link the world, or at least vast regions of the world together, become simultaneously places where these connections become vulnerable to attack or breakdown. Electric blackout phenomena can only occur because of vast, complex network systems which, when one node fails, can 'crash' an entire regional system; and air travel which connects peoples from all continents becomes vulnerable to threats of home-mixed explosives carried by terrorist individuals which when detected or even suspected can close down entire international travel systems.

5. With respect to present and future inquiry, how can the most important philosophical problems concerning technology be identified and explored?

The philosophy of technology remains a relatively young subdiscipline in philosophy. Although there were late 19th century intimations with Karl Marx and Ernst Kapp, its origins are largely 20th century. Even the term, "technology," itself is a largely 20th century term and use. David Nye, a historian of technology notes that there are very few references to technology in the late 19th century, with "inventions" and "applied arts" being more common until after the first World War. And, it is just after the first World War that philosophers, particularly in Europe, began to use the term, "technology," and to write book-length works about or including discussions of technology. Martin Heidegger's discussion in *Being and Time* and Friedrich Dessauer's *Philosophie der Technik* were both published in 1927. Of course, as contemporary historians of the philosophy of technology now note, while many major European philosophers began to talk about technology, most could be characterized as dystopian; as dealing with technology-in-general; and as taking technology as a threat to many perceived values of traditional culture. The one major exception to this tendency was the American philosopher John Dewey, who saw technologies and technological thinking as an instrumental means for social improvement and the dissemination of democracy. However, by the late 20th century, Hans Achterhuis and a group of Dutch philosophers of technology discerned what they termed "an empirical turn" amongst now mostly American philosophers of technology. In a book, *American Philosophy of Technology: The Empirical Turn* , they analyze six contemporary philosophers of technology, all of whom do concrete or empirical studies of types of technologies, take a pragmatic attitude, and reject the more 'transcendental' or metaphysical attitudes of the mid-century European forebears. And, since then, an even younger generation of philosophers of technology, have followed this style of analysis into yet many other regions of technological development. I am not only sympathetic to such an 'empirical turn,' but think this is how philosophical problems concerning technologies can and should be identified and explored. At the same time,

I also believe that it is important for philosophers in particular to immerse themselves in 'histories.' Here is an example of why both concrete and historically informed inquiries are needed: Currently there is a strong set of debates raging concerning both privacy and freedom of expression relating to telephonic and internet communications. The revelation that the U.S. Government has been eavesdropping on these media without the usual protection of court approvals, has raised again questions of the freedom of speech and freedom from unwarranted governmental searches. Yet, those who know something about the history of freedom of speech alongside the histories of media, will recognize that this is a battle which re-appears with each new medium. The very long fight to secure freedom of the press—for print media—has been followed for each different medium, for example, radio – and television – and now, the internet. Involved in these debates are arguments about how each new medium is 'different' and thus must be controlled or not controlled according to its difference. These issues are also of importance philosophically. But they are contexted in the particularities of technologies and the histories of uses. I earlier did attempt to make some observations about larger patterns related to ensembles of technologies and contemporary problems related to technological developments. In *Technology and the Lifeworld* (1990), I argued that contemporary technologies do produce certain changes in the way individuals and societies respond, including increased decisional burdens, by which I meant that we increasingly must consciously decide things which in other times and societies did not call for explicit decisions—living wills, birth control, organ use, are all symptomatic of such results. Similarly, what I called oscillatory phenomena, are also common. Immediate information circulation and very fast travel have made responses to crises, wars, natural disasters, much more instantaneous and much more global. International student revolts in 1968 were an early example; but more recently the Indian Ocean tsunami and Pakistan earthquake were other examples of virtually immediate and global responses. Such phenomena could not take place without contemporary communications and transportation technologies, but also take the particular shapes they do within the constraints and capacities of such technologies. Note, for instance, what kinds of differences occur with respect to the destruction of roads, bridges, and the like in earthquake situations, overcome today by the helicopter, yet the very shortage of helicopters became an issue in Pakistan. We are here back again to the previous question and the history of ideas—how can we interpret our situations today unless we include and emphasize technological issues therein? Finally, I will return to some specific issues with which I have become concerned in recent years: from my early interest in philosophy of technology, I have been particularly interested in the role of technologies in the production of knowledge and, with respect to science, the role of instrumentation. This has led me in several directions including, currently, a now decade long study of imaging technologies. And this investigation has led to some rather interesting conclusions along the way. For example, if good science includes measuring perceptions, that is observations aided by some standardizing instrument, then in a minimalist sense the most ancient and the largest variety of cultures or peoples have produced much science from astronomy. Virtually all ancient peoples learned to recognize seasons, the solstices, had calendars and accounted for celestial cycles—and it can be shown that instruments were part of this system of observations, from Stonehenge to sun daggers to architecturally contained observation tubes. Early modern science, at least in Europe, was based in many cases upon lens technologies, particularly telescopes and microscopes and with now magnified perceptions began to experience a world much larger and much smaller than in previous histories. By the time one reaches late modernity, imaging technologies, detect and image phenomena far exceeding our bodily perceptual capacities, but through imaging technological mediations, images are translated back into visible shapes and configurations which are perceivable, again expanding our worlds and producing knowledge through instrumentation. Indeed, late modern science without instruments would have to shrink to the limits of a now ancient past. Here, again, is a role for philosophy of technology, a philosophy sensitive to the material culture, conditions and capacities of human-technology relations.

Selected Bibliography

Books

Hermeneutic Phenomenology: The Philosophy of Paul Ricoeur. Evanston: Northwestern University Press, 1971.

Sense and Significance. New York: Humanities Press, 1973.

Listening and Voice: A Phenomenology of Sound. Athens, Ohio: Ohio University Press, 1976.

Experimental Phenomenology. New York: G.P Putnam's Sons, 1977; Albany: SUNY Press, 1986.

Technics and Praxis: A Philosophy of Technology. Dordrecht: Reidel, 1979.

Existential Technics. Albany: SUNY Press, 1983.

Consequence of Phenomenology. Albany: SUNY Press, 1986.

Technology and the Lifeworld. Bloomington: Indiana University Press, 1990.

Instrumental Realism: The Interface between Philosophy of Technology and Philosophy of Science. Bloomington: Indiana University Press, 1990.

Philosophy of Technology: An Introduction. New York: Paragon House, 1993.

Postphenomenology: Essays in the Postmodern Context. Evanston: Northwestern University Press, 1993.

Expanding Hermeneutics: Visualism in Science. Evanston: Northwestern University Press, 1998.

Bodies in Technology. Electronic Mediations Series. Volume V. Minneapolis: University of Minnesota Press, 2002.

Monograph

On Nonfoundational Phenomenology. Publikationer fran institutionen for pedegogik, Fenomenografiska notiser 3 (Goteborgs, 1986).

Books Edited

Coeditor, with Richard M. Zaner. *Phenomenology And Existentialism.* New York: Capricorn Books, 1973.

Editor, Paul Ricoeur, *The Conflict of Interpretations.* Evanston: Northwestern University Press, 1974.

Coeditor, with Richard M. Zaner. *Dialogues in Phenomenology. Selected Studies in Phenomenology and Existential Philosophy. Volume V.* The Hague: Martinus Nijhoff, 1977.

Coeditor, with Richard M. Zaner. *Interdisciplinary Phenomenology. Selected Studies in Phenomenology and Existential Philosophy. Volume VI.* The Hague: Martinus Nijhoff, 1977.

Coeditor, with Hugh J. Silverman. *Hermeneutics and Deconstruction. Selected Studies in Phenomenology and Existential Philosophy. Volume IX.* Albany: SUNY Press, 1985.

Coeditor, with Hugh J. Silverman. *Descriptions. Selected Studies in Phenomenology and Existential Philosophy. Volume X.* Albany: SUNY Press, 1985.

Coeditor, with Evan Selinger. *Chasing Technoscience: Matrix of Materiality.* Bloomington: Indiana University Press, 2003.

13

Ian C. Jarvie

Distinguished Research Professor of Philosophy, Emeritus

York University, Toronto
Canada

---

1. Why were you initially drawn to philosophical issues concerning technology?

The answer is that my interest was aroused in the course of my education. In his LSE introductory lectures on logic and scientific method Popper analysed the logic of explanation as deductive in what is now a standard way:

Universal statements do the explaining together with Factual statements (initial conditions).

Which jointly enable the deduction of a Statement of the problematic facts.

To explain the problematic fact of a solar eclipse it was necessary to have a set of theories about celestial mechanics (perhaps plus optics), plus a set of factual statements about this particular planetary system, and how it looked to observers on the earth at different points and times. Together these would be sufficient to deduce, i.e. predict, a solar eclipse at a particular point on earth. This assimilation of explanation and prediction to deduction really made me sit up. (Arguments by critics that point to occasions when we use explanation and prediction in ways that cannot be assimilated to deduction strike me as less than persuasive.) A second example relevant to social science made me sit up even straighter. Consider the fate of King Charles I of England.

All men die when their heads are cut off.
King Charles I had his head cut off.

King Charles I died.

The example is not very satisfactory to the historian, since when the historian asks why the English king died what is being requested is not the conditions under which human beings expire. Similarly, the example is not very interesting to the biologist, since the generalization with which it begins, while likely true, is just another fact that needs explaining.

Popper then showed that if one's aim was to find out why human beings die upon being decapitated one's interest might be described as scientific. One was interested in the universal aspects of the matter. If one's aim was in the circumstances of how it came about that the English monarch was decapitated that interest was historical, and quite particular. If one's aim was to devise a humane method of executing human beings, as was said of M. Guillotin, then one's interest was applied or technological. There were, then, generalizing or theoretical sciences, such as physics and sociology; there were historical sciences, such as history or geology; and there were applied sciences, such as engineering, including social engineering. Explanations in both historical and applied sciences drew upon the results of theoretical or generalizing sciences which were thus logically more fundamental.

I may have embroidered or reordered matters in memory, but this is what I recollect of my first encounter with the idea that the sciences and technology were distinct in a logical and not just in a practical way. (The treatment given to these matters in Popper's *The Open Society and Its Enemies* is not quite the same. See Popper 1945, chs. 5, sec. IV and 25, sec. II.) Good engineering, including good social engineering, would be that premissed on the way things actually were, as described in our best approximation to it, namely current science. Hence theoretical study of society was not a dispensable activity but quite essential background to effective and responsible social reform. If one disputed some policy measure or other a preliminary to debating it was to refer to our scientific knowledge, such as it was. This was a controversial position since there was then as now considerable scepticism as to whether there were any social science results of the kind that could lead to reliable social technology. Popper did not endorse the demand for reliability. His points were that all science was conjectural, and that the correction of error was a deductive process.

This was the same in a so-called "Humanity" like history, as in hard-nosed physical science. Technology was part and parcel of science so understood, and was deductive and conjectural like the rest. Erroneous technological conjectures were not purely naturalistic, they had a moral and social dimension built in to their very formulation. Effective and responsible technology cried out for good natural science, good social science, and a responsible moral attitude.

Taking this insight under advisement, I had for long thought that we philosophers of science took insufficient interest in technology. I debated with my friend and colleague Agassi his ideas that a further distinction was needed between applied science and technology, given that one could attribute the scientific aim to the former but hardly to the latter. Hence I eagerly accepted Melvin Kranzberg's invitation to participate in a symposium on the philosophy of technology held at the joint meeting of the Society for the History of Technology and AAAS at San Francisco in December 1965. The symposium was published in *Technology and Culture* the following year.

2. What does your work reveal about technology that other academics, citizens, or engineers typically fail to appreciate?

Two points have been my main concern and, I think, contribution. First, that technological problems are constructed or specified by society and its agents, not by technologists. Second, that the demarcation between the natural and technological (or man-made) is not one that is easy to draw.

To expand on my first point. A typical engineering problem would be to design a bridge across a ravine. It looks at first glance to be a problem in materials suitable to the length, height, soil, rock and climatic conditions of the place where the bridge is to be built. These in turn are a function of the weight and kinds of traffic it is to bear. An engineer does not need a philosopher to tell her that there is no perfect solution, no bridge that is guaranteed never to fall down. The engineer works within given parameters. Some of these are material, such as just what materials are available and what is known of their strengths and weaknesses. But the questions are set socially. First and most obviously, there is budget. The budget allocated for the project is not one determined by the engineers. It is, rather, determined by the politicians and bureaucrats whose responsibility is budgeting. Engineering input is one among many factors they must weigh. More significant, there are the administrative and legal safety standards that must be met. These may be drafted by engineers, but they are enacted by legislators or their equivalent. Some projects are such that there are no agreed standards to be met. A case in point is what was revealed after Hurricane Katrina, about the levees, locks, and canals intended to protect New Orleans from flooding. What was revealed was a social background of debate about how to use the natural protections afforded by the delta marshlands, what standard to set for the constructions, for example whether they were to be adequate to resist a Category 5 storm, and the very far from free hand the Army Corps of Engineers had due to this background and to budgetary constraints. Not to mention that some of the work was improvisatory. No one doubts that protections as effective as those of the sea walls of the Netherlands could be built, that is, that the engineering expertise exists. But engineering problems are given to the engineers surrounded by a host of social constraints and specifications within which they must work.

To expand on my second point. Few distinctions are as basic as that between nature, the given, and conventional, or man-made. The one is what cannot be otherwise; the other is what humanity has wrought. Yet when one examines the centre of the distinction, the human being, natural and man-made are inextricably entangled. The human being is engineered: nutritionally, educationally, by socialization, by surgical intervention, by medical treatment, and so on. Clothes maketh the man, it is said. Humans endlessly fiddle with their bodily hair; pierce, cut, change, and colour their flesh; adjust their posture; restrict their food intake; and so on. They ensure that their groups are differentiated. They treat with their fellows in numerous ways that both endorse and deny their commonality. Homo sapiens is nowhere in a state of nature, and never was, pace Rousseau. There is nothing corresponding to innate human nature. So, if we cannot pinpoint a natural core, we cannot lay down a demarcation. We may know that there were oscillating ice ages and thaws in the time before human beings strode the earth. Now that they have been around for a long time we may never be able to disentangle whether future oscillations, or lack of them, are not partly man-made and not merely natural. This interweaving is something we are stuck with, neither to be lamented or celebrated. It is the problematic inside of which we think.

It affects some technology profoundly. Not only are the problems socially set, but the materials to be used are a mix of the natural and the man-made and will not necessarily behave in conformity with the slide rule. No doubt working engineers take account of much of this in their work. Whether they make it clear enough to the public that is ultimately setting their agenda is debatable.

3. What, if any, practical and/or social-political obligations follow from studying technology from a philosophical perspective?

What follows is the realisation that engineers are social engineers, like it or not, and that humans are "naturally" engineers of their surroundings, which simply ties together even more tightly the issues of what humans do to their surroundings and their responsibilities to themselves and others, including those not yet born. More probingly, it casts expertise and the authority of expertise in a different light. The naïve view is that the authority of expertise derives from the knowledge the expert possesses. But when the expert belongs to a licensed monopoly of experts it can be argued that their authority derives not from their knowledge but from the social and political system that created the monopoly. The naïve view is a mystification and leads to disappointed expectations when expertise fails to deliver the goods. Licensing monopolies of experts creates a situation where there is bound to be a conflict of interest between offering the expertise in a rational and critical manner on the one hand, and defence of the profession on the other. These clashes can lead to lawsuits and public inquiries. It is hardly good social planning to create a system of built in conflicts of interest. But debate about improving the system is a public rather than a specialist responsibility.

4. If the history of ideas were to be narrated in such a way as to emphasize technological issues, how would that narrative differ from traditional accounts?

Point one, I am not an historian of ideas or of technology, hence it would be hard to assess difference. Point two, more fundamental, I am leery of questions in the subjunctive mood, viz. "were to be narrated". The truth conditions of such conditionals are highly controversial.

5. With respect to present and future inquiry, how can the most important philosophical problems concerning technology be identified and explored?

According to my account, mentioned above under 2, this is a question that falls out from the technological problems that the society settles upon. The major contributions that philosophers can make is to insist and continue to insist upon the basic point that technology should not be confused with science, and that technological problems are a social not an individual responsibility. To take the well known example of atomic energy and the atomic bomb, these technologies were developed first in a military context and only subsequently in a civilian one. So long as the defining social problems were military the wider social impact was shoved to the background. But we know from the Bulletin of the Atomic Scientists that even as the feasibility of a chain reaction became known some of the scientists could see at once that the social consequences were immense and immensely dangerous. They were supposed to focus on the technical challenges. They saw vividly that the licence to pursue technological research was not adequately left to the military, was not safely kept secret from the public at large. Consequences so grave, financed by the public, needed to be entered into with knowledge and consent. Yet it was and remains a case where a substantial body of informed opinion maintains the anti-democratic position that the public has no right to know, and indeed should be protected from knowing what is done in its name and with its revenues.

The most important philosophical problems of technology are, then, social and political ones. Neither in civilian use, nor in military and clandestine use, is there remotely adequate input from the general public about technology, its costs and its benefits. The general public are the ones who will suffer from their ignorance and their trust in the responsibility of their leaders. The public delegates technological monopolies to engineers, doctors, lawyers, pharmacists, and so on. If these perform unsatisfactorily it is vital to realise that these monopolies are creatures of democratic states and societies so that the responsibility for fixing them lies not only with the practitioners. In general we are insufficiently proactive when it comes to our democratic responsibilities. We continue to elect political leaders who regularly insult our intelligence and condescend to us. What is needed is better social technology to overcome these supervisory deficits.

Selected Bibliography

Jarvie, I. C. 1966 "The Social Character of Technological Problems", *Technology and Culture, Vol.* 7, July, pp. 384-90. Reprinted in Carl Mitcham and Robert Mackey, eds., *Philosophy and Technology*, New York: Free Press, 1972, pp. 50-53; also in F. Rapp, ed., *Contributions to a Philosophy of Technology: Studies in the Structure of Thinking in the Technological Sciences* (*Theory and Decision Library 5*), Dordrecht: D. Reidel, 1974, pp. 86-92.

—1967a "Is Technology Unnatural?", *The Listener, Vol. 77*, March 9th, pp. 322-23 and 333.

—1967b "Technology and the Structure of Knowledge", the banquet lecture to the annual conference of educators in technology, delivered at, and printed by, the *Division of Industrial Engineering and Technology*, State University College, Oswego, New York. Reprinted in Carl Mitcham and Robert Mackey, *Philosophy and Technology*, New York: Free Press, 1972, pp. 54-61.

—2001 *The Republic of Science. The Emergence of Popper's Social View of Science, 1935-1945.* Amsterdam: Rodopi.

Popper, K. R. 1945 *The Open Society and Its Enemies.* London: George Routledge and Sons.

14

Bruno Latour

Professor, Centre de sociologie de l'Innovation

Ecole nationale supérieure des mines, Paris
France

1. Why were you initially drawn to philosophical issues concerning technology?

I am not sure I ever was ... Or rather, I am writing this after twenty-five years in an engineering school—the Ecole des Mines, a French grande école where the elite engineering bodies of France are supposed to be trained. And in a way, what I feel now is my largely failed attempt, over this quarter of the century, to practice the philosophy of technology. So I am afraid what I am going to say may sound a bit self-critical: What I want to understand is why I have failed so utterly. But first, let me make a point about vocabulary: In France, we understand technology as the philosophy, or the reflection, or the science about techniques, in the same way as epistemology is the reflection about science, the science of science if you wish. No one will say about a new particle found at CERN that "it is an epistemology". There is no more reason to say about the newest culinary robot that it is a "technology". It is a technique, to which might be added or not (most often not) a reflection by some scholar.

I insist on this usage because the reason why there is so little philosophy of technology is because it is always thought in relation, or under the shadow, or in the dependence of the philosophy of science. The worst philosophy has been done by people using the word "technoscience" as if the two were the same domain. (I have used the terms however in *Science in Action* and very much regret it.)The extreme case being of course Heidegger whose point is to fuse the two inside the notions of domination and thing—thing being conceived as the ultimately mathematized entity. If there are two mistakes not to commit when dealing with techniques, it is to think them as "technologies", that is as "applied science", or as a sub-case of mathematized objects—and of course to take them as a case of domination "of Man over Matter" as the cliché goes. There is not that much matter to begin with. As to domination ... you really need to be pretty ignorant of techniques to think of them that way. So the best solution to maintain that distance and to shake the weight that epistemology exerts on technology (conceived as the study of techniques) is to keep the word "techniques".

Now to answer your question, I was drawn to technology precisely because of my diffidence for epistemology –learned in science studies – and also because I have lived among engineers since 1979, first at the Conservatoire des arts et métiers – a revolutionary invention and still a marvellous museum of techniques – and then at les Mines. What struck me from day one, was how different techniques were from science and how ill-equipped we were in science studies to deal with issues of technical studies (even though we kept using the same acronyms STS or S&TS). But I have to say that we have in France a very different philosophical tradition which has the great advantage of foregrounding the originality of techniques quite a bit. Apart from Ellul, a moralist who believes that techniques are what epistemologists say they are (domination over matter and over humans), we have a very rich tradition from Diderot, Laffite, Bergson, André Leroi-Gourhan all the way to François Dagognet, a lesser known but quite interesting figure, andCornelius Castoriadis. If you read Leroi-Gourhan, it is not philosophy, but it is technology at its best.[1] And of course Gilbert Simondon, whose book remains one of the few from a philosopher to pay respect to the complete originality of techniques.[2]

I remind you that in his book Simondon connects techniques to a genealogy – he calls it a "genetic", and it is mythical of course – that has the mode of existence of techniques emerging out of magic together with religion, and only later giving birth to science and morality with the arts doing the mediation and philosophy the synthesis. Quite amazing! His whole idea is that you don't understand techniques if you don't understand magic and religion and the ways forms and background are distributed.

All of this to say that for a French philosopher entering engineer-

---

[1] Leroi-Gourhan, A. (1993). *Gesture and Speech.* Cambridge, Mass, MIT Press.

[2] Simondon, G. (1958). *Du Mode d'existence des objets techniques.* Paris, Aubier.

ing school with a science studies background, it was not surprising that the complete originality of techniques would have struck me enormously. And of course, when I entered the field, Michel Callon had already started his pioneering study of the electric vehicle. So I had no real merit. Then I met Wiebe Bijker and the whole thing developed. Don MacKenzie published his masterpiece on guidance system.[3] We connected with Tom Hughes and the historians of technology. And of course I wrote *Aramis or the love of technology* – to this day, my favourite book – and I was drawn into it. Is this what you would call philosophy of technology? I am not sure. I believe in philosophy, but not in philosophy "of" something, so I am not sure I have contributed anything to that field.

2. What does your work reveal about technology that other academics, citizens, or engineers typically fail to appreciate?

Well, you should ask them for that, not me! ... If there is one thing a writer is unable to say it is what his work "reveals", because to reveal you need to be two and even three: you, what you are talking about, and who you are talking to. You can say "look here! how wonderful!", but if the one you address is looking the other way, you can be marvelously perceptive, no one else will notice... My general feeling is that when I have talked about the importance of techniques, people looked the other way, except if they are fans of that very narrowly defined technique, like railway people, or plane buffs, or computer wanks, etc. In other words, there is no general conversation about techniques as such, either it is very specific or when it is generalized it is taken over by epistemology and then you have empty clichés about "technology overpowering its masters" etc. People have not even been able to quote the Frankenstein's myth faithfully ... When you talk about science and epistemology, people at least notice because they have attached so much morality and politics to them that any change in the theory of science makes them reach for their guns, but not to techniques?! You will notice that there has been a "science war" (a pretty silly one), but no "techniques war". Maybe it is just my experience, but I have not met much success.

[3] MacKenzie, D. (1990). *Inventing Accuracy. A Historical Sociology of Nuclear Missile Guidance.* Cambridge Mass, MIT Press.

Anyway you cannot lump all of those groups together, and you have to differentiate academics, engineers and citizens. Academics, as a rule fail to appreciate so many things, that it is hard to know where to start! There is this near impossibility with modernism and modernists in general to be sensitive to what is given in experience that baffles me. There are still people who fret in sociology, anthropology and maybe philosophy, because in my definition of techniques "I give a role to non-humans" ... and they pronounce this sentence as if they were saying "Latour is a pervert, a zoophile" or something of the sort. We have been connected, attached, folded with non-humans for millions of years, and especially for the last three centuries, and it would come as a surprise for academics?! How strange. In my experience, academics live in a world that still predates all the industrial and technical revolutions. They are sort of upper Paleolithic – and even that is unfair because in that time they had already lots of stones ... and when you see the way philosophers treat stones, it is not encouraging ... So what I can "reveal" to them is non-existing, since they see it as a revolting promiscuity or a "coquetterie" of my part. Luc Boltanski, the best sociologist in France, still believes that "all this talk about non-humans" in my sociology is a pose, a way I have found to render me interesting; he really believes that I can't be serious. For him, sociology is for humans only, any other thought is simply, as he put it, "poetism"; he told it to me again last week in a defense thesis where one of my students has been studying how some French anglers became militants for the quality of rivers: Their entire life revolves around fish, and their whole politics depend on this conversion of passions for angling into a new passion for defending the quality of the rivers. But the best French sociologist still wants my student to swear that she will abstain from this silly sociology (or philosophy) of mine! What can I say? "Poetism" means that if people speak in a way that seems to link humans and non-humans in strange ways, they don't mean it, it is simply metaphorical, poetical, false ... (Which is quite funny because Boltanski is also a poet in his spare time.) So the bottom line is that years after the dispute with Collins and Yearley, everything is still the same: Academics seem to reply to my work, "Please don't take seriously what people say about techniques and material relations, they just don't know what they are saying, ignore what they say or do and keep separating the two as Descartes told us to do".

Now, engineers. Here I have failed so utterly that I did not really want to answer your questionnaire for that reason. I have been in les Mines for twenty five years, and not once did I interest even one minute any of the engineering professors there. Did I try? Yes. How good was I at convincing them? I have been probably pretty bad. The best I could get as an effect after 25 years was that some professors condescended to consider that "in addition to the technical aspects" of, I don't know, mining, statistics, robotics, informatics etc., there might "also be some aspects" to be "taken into consideration": "yes, there are also social elements".... which, for them, usually means at best "acceptability" by the public. And I don't think Callon did any better in that school. The students have been interested in my class on "Mapping scientific and technical controversies", but not once was it seen as an important aspect or as a way to renew engineering culture in France – even though everyone is complaining about the "rise of irrationality" and the "decline of enrolment in scientific careers". I am actually fascinated by this failure because now that I am leaving to go to a social science school, Sciences-Po, a sort of French LSE, I am planning to do the opposite: that is, to interest politologists and historians in techniques. I will probably fail just as well but I want to try. So as far as engineers are concerned, the conclusion is that none of the work I did has revealed anything of significance to them about what they do, not even a new positive version of how interesting they are: I keep telling them "how interesting you are because you connect humans and non-humans in so fascinating imbroglios and so deep and opaque labyrinthic practice" and they answer "no, thanks, we are just plain boring, would you please stop being interested in us ... "!

Citizens are another matter. I think that with *Politics of Nature* and with *Making Things Public* I have entered into quite a lot of interesting conversations with those who realize that any new techniques are an assembly of some sort for which the representative parties are not "constitutionally" gathered. In that sense, I think I had a slightly better reception among journalists, politicians, ecologists; but again, this is marginal. The pretension of revealing anything to others is always empty anyway. People pick up whatever they want to pick. What I have done is to propose a rather rich and, in my view, tasty dish for which people can chose if they wish to come to the dinner table. But if they prefer modernism, what can I do? The problem with techniques are that people love to hate them and also hate to love them, no matter if they are academics or not, so it is extraordinary difficult to get the right distance with the mass of thing which they cohabit.

3. What, if any, practical and/or social-political obligations follow from studying technology from a philosophical perspective?

I am not sure again of what you mean by a "philosophical perspective". I only know how to study a subject matter by trying to be faithful to what is given in experience: That can be done only by description and, in the case of techniques, thick descriptions—given our utter ignorance and the rarity of their registration in culture and humanities. If by philosophy you would mean an attempt to add to those descriptions "foundation", "reflexiveness", "transcendental" principles, I really hope to have no philosophical perspective on anything, least of all on techniques. I am an empirical philosopher, and what I try to do with techniques is, as Simondon, urged us to do, to detect what is unique in their mode of existence. As long as this uniqueness is not detected, honored, celebrated and cherished, there is no way to call us humanists or to say that we teach "the humanities". So for me, most academic life, most literature, most humanities are deeply barbarians: They ignore, despise, love to hate, and hate to love what make us humans. This is why I think Richard Powers' enterprise in literature is so important: he has done in my view infinitely more for philosophy of techniques than any of us.

Does this have any practical or socio-political implications? I suppose. First, it has teaching implications: I have tried to do that for years, displaying the connections and the opacity of techniques and the importance of technology as a mode of existence, which could help students to reconnect with them instead of always cutting the connections and saying "they are just mere objects" – as the Carthaginians were doing when throwing their children in sacrifice to Baal and claiming they were animals ... Second, the politics of the future depend entirely on the careful and complex maintenance of involment with things whose ecosystems will be even more fragile and multiple. To dream of mastery and domination "over matter" is simply silly. But this is the politics of things, the Parliament of things, as I said, that I have tried to envision and then to scale model in *Making Things Public*. It has obligations, in Stengers' sense; it imposes obligations on me at least.

Can I say that it has practical implications? It is very hard to say. I have done lots of work, in the past, in the management of innovations, I am not sure it had any impact. Do philosophers have impact in the end? Yes and no. What we do is too marginal to be evaluated by its impact, but may be I failed to understand the question. I really feel that for thirty years I have behaved as if I was talking to non-existing people, those who would have accepted that we have never been modern, but as long as they believe to be modernists or post-modernists, I think it is fair to say that my work is fairly useless. And in addition what does it mean to have "socio-political obligations"? If those obligations are defined by Isabelle Stengers or Ulrich Beck they are pretty different from those defined by Steve Fuller ...

4. If the history of ideas were to be narrated in such a way as to emphasize technological issues, how would that narrative differ from traditional accounts?

I think it is in science studies that notions like practice, know how, space, equipment, innovation, laboratory networks, actants and so on, are used. In a sense, it is fair to say that techniques have become the way to understand *episteme*, and in that sense, one could say that science studies have largely counteracted the original prejudice of philosophy which rendered techniques subservient to science. Most of the new social history of science – I am not talking of history of ideas since it has disappeared from view and rightly so – is inspired and in the shadow of an attention to techniques in all of the meanings of the word: one of them being art, another being know-how, a third being "intellectual techniques" or "paper techniques" and of course the most important being the new role given to instruments. It is also the trend of the sciences themselves because of their heavier, costlier and more visible materiality. So they tend to offer themselves more to the grasp of technology than of epistemology. Now it is obvious when people talk about materiality, that it bears no relation whatsoever to what was meant by it in late modernity. I don't think a major book like Reviel Netz on mathematical practice in Greece would be possible without a disclosure of what we mean by techniques.[4]

What has always interested me, of course, is how the grand narrative of the human race would be modified if we were to include the non-humans in it in a productive way and not just as "matter".

[4] Netz, R. (2003). *The Shaping of Deduction in Greek Mathematics : A Study in Cognitive History.* Cambridge, Cambridge University Press.

Or rather what would happen to materialism and "the domination of Man over matter" if we had at last a realistic definition of "matter" like the one inspired by technology. It is because techniques are never about mastery, domination, prediction, but always about surprise, ruse, detour, opacity, obscurity, arrangement, attention, care, complexity, unintended consequences, translations, folding, and labyrinth that they have interested me so much. You find this complete originality of techniques any time you pick up a book on one technique, on steam engines, on pencil, on paper clips, on accounting, on diagrams and so on. The last one I read was on the invention of the "container" and how this box "changed the world". It is impossible to talk about globalization without taking the contained into account.[5]

Now you have hundreds of books on how a given technique "changed the world", but we have no master narrative of what it does to our representation of ourselves to be technical through and through. I think it would be really interesting to try. I made a feeble attempt in *Pandora's Hope* and in my work with Shirley Strum, but we have not yet escaped the simplistic "materiality" or the Marxian narratives with the *Homo faber* myth. In that vein, anthropologists like Philippe Descola, Marshall Sahlins, Tim Ingold, probably could teach us much more on how to liberate matters from the silly materialism of the past – which was nothing in fact but a simplistic projection of geometry by epistemologists onto technical questions. Or even more simply, an artifact of projective geometry, a confusion between drawing a machine on paper, as if the mode of existence of a technique had any direct relation with the ways they are drawn! I have been fascinated by this question and I hope to make it the topic of my next exhibition and catalogue: How come modernists have ended up imagining that matter – the matter of their 'materialistic', 'mundane', 'down to earth' mythology of efficiency and reductionism – could be a piece of paper? How could they have confused *res extensa* and materiality when every object, every artisan, every skilled gesture would have told them the very opposite? In present philosophy the thinker who is doing most to reinvent a master-narrative that factors techniques in, is certainly Peter Sloterdijk in his three volume *Spheres* story. It is not always done with the care I would

[5] Levinson, M. (2006). *The Box: How the Shipping Container Made the World Smaller and the World Economy Bigger*. Princeton, Princeton University Press.

like to see, but it is a master narrative of humanity born out of techniques and I am all for it.

5. With respect to present and future inquiry, how can the most important philosophical problems concerning technology be identified and explored?

Philosophy does not proceed with research programs and grant applications, but one can still identify some topics for which we are in great need of meetings and descriptions. One of course is the very notion of non-humans. How can we talk about "tools" in a not "toolish" and thus "foolish" way? Graham Harman, a young philosopher of technology, has given a lot of thought to that question in redescribing Heidegger's concepts. I don't believe in a subfield of philosophy called philosophy "of" technology, there is only philosophy, but in proper metaphysics the ontology of tools remains a complete mystery.[6]

Another topic is to reassess the origin story proposed by Simondon: to link it to magic and to make it a *vis à vis* of religion is odd, but it is probably a very strong inducement to try to anthropologise even further the connections humans entertain with non-humans. Another very promising growth area is in reaching out to anthropologists: now that the master book of Descola offers a map to chart the human non-human connections in many different collectives, we might reconsider the whole notion of material culture and in a way rewrite the entire Marxism without having to fall back into the Efficiency, Objectivity, and Profitability mythology.[7] I am thinking here of the fabulous book on *Barbed Wire* by Reviel Netz (yes the same author as the other masterpiece!) probably one of the best "materialist" studies I have read since Cronon's book about Chicago.[8] A materialism that will import in its definition of materiality the insights we all had about technology would make a big difference.

But none of that will do if we are not able to invent *visualizing* procedures to render vivid to the eyes of those we try to address the new shape of techniques. The problem as I see it is

---

[6]Harman, G. (2002). *Tool-Being: Heidegger and the Metaphysics of Objects*, Open Court.

[7]Descola, P. (2005). *Par delà nature et culture.* Paris, Gallimard.

[8]Netz, R. (2004). *Barbed Wire: An Ecology of Modernity.* Wesleyan University Press; Cronon, W. (1991). *Nature's Metropolis. Chicago and the Great West.* New York, Norton.

that we are still dramatizing techniques with the vocabulary invented in the Renaissance to invent and collect and draw them together. In other words, the optical space in which techniques enter the world of humanities – from childrens books to CAD design computer screen to technical museums – is still that of the Quattrocento: immutable mobiles and exploded views plus more than a few hypes and in the end some moralization. If you want to portray them "realistically" as I tried to do with *Aramis*, that is, to show them not as an object but as a project, as a "thing", as a "collective", as an *assembly of assemblages*, there is no good way and we are left with only the fragile web of words. I don't think we will convince any one, and be able to teach, as long as we don't benefit from an enterprise that should have the magnitude of the Quattrocento invention of perspective. Except, of course, it is not perspective: more like what Peter Sloterdijk calls "spherology", the study of technospace, of bubbles and envelops. Can philosophers do it? Alone not. But with artists, historians, computer scientists, designers, citizen groups, maybe. Again can we "draw things together", can we make "things public"? These are the questions that are still attracting me to technology and to which I hope to be able to contribute a bit, even though I have so thoroughly failed to interest my colleagues in the engineering school where I have developed all of those ideas. Maybe philosophers always write for absent people.

Selected Bibliography

Latour, B. and S. Woolgar (1979 1986). *Laboratory Life. The Construction of Scientific Facts* (second edition with a new postword). Princeton, Princeton University Press.

Latour, B. (1987). Science In Action. How to Follow Scientists and Engineers through Society. Cambridge Mass, Harvard University Press.

Latour, B. (1988). *The Pasteurization of France.* Cambridge Mass., Harvard University Press.

Latour, B. (1993). *We Have Never Been Modern.* Cambridge, Mass, Harvard University Press.

Latour, B. (1996). *Aramis or the Love of Technology.* Cambridge, Mass, Harvard University Press.

Latour, B. (1999). *Pandora's Hope. Essays on the reality of science studies.* Cambridge, Mass, Harvard University Press.

Latour, B. (2004). *Politics of Nature: How to Bring the Sciences into Democracy* (translated by Catherine Porter). Cambridge, Mass, Harvard University Press.

Latour, B. and P. Weibel, Eds. (2002). *Iconoclash. Beyond the Image Wars in Science*, Religion and Art. Cambridge, Mass, MIT Press.

Latour, B. and P. Weibel, Eds. (2005). *Making Things Public. Atmospheres of Democracy.* Cambridge, Mass, MIT Press.

15

Bill McKibben

Scholar in Residence in Environmental Studies

Middlebury College
USA

---

1. Why were you initially drawn to philosophical issues concerning technology?

Because of my background as a journalist I'm interested in the biggest stories—the ones that have the power to redefine the world. So I wrote the first book for a general audience about global warming, and then, more recently, one of the first accounts of what we might be facing with germline genetic engineering, advanced robotics, and similar technologies. But what interests me is less the depiction of the possibilities—though that's a necessary part of my work, and one I enjoy—than an attempt to figure out what they mean to our understanding of who we are and how we fit into the world. Hence, with global warming, *The End of Nature* focused on what it will feel like to live in a world where humans influence everything around them. *Enough* tried to grapple with what seems to me the fragility of human meaning.

2. What does your work reveal about technology that other academics, citizens, or engineers typically fail to appreciate?

That change – radical change – may be bearing down on us. It's very hard for us to understand that. Even if we, say, believe in global warming we think it must be somewhat distant, happening fairly slowly; this distortion, probably due to our evolved mechanisms for assessing risk, makes our reactions slow and limp. Similarly, a discussion of the possibility of, say, designer babies sounds like science fiction to most people, never mind the fact that we've done the same thing with any number of plants and animals and that there are plenty of scientists who want to get going. It's very hard for us to perceive that change can be discontinuous, sudden, and enormous. It's very difficult for us to imagine we might be standing at a strange moment in human history, right on a threshold.

3. What, if any, practical and/or social-political obligations follow from studying technology from a philosophical perspective?

I can only speak for myself; once you've thought about such issues for a while, the need to take action becomes self-evident. I've spent much of this year, for instance, helping organize campaigns to force political action on global warming—a fifty-mile march I helped lead across New England became the largest demonstration yet on climate change in this country, and caused several of our federal candidates to switch their position on the issue. I find that most people are motivated more by the philosophical and moral fears about such issues than by more practical considerations, which I think is very good—it's healthy, I think, to register a visceral unease with the prospect of designing children in embryo to be 'better.'

4. If the history of ideas were to be narrated in such a way as to emphasize technological issues, how would that narrative differ from traditional accounts?

I don't know much about the history of ideas in an academic sense, but I do know that technological change has driven so many of the ways we understand the world. For instance, the signal moment of the modern age came when we figured out how to combust fossil fuels and thus dramatically expand our powers. This led to a thousand other things—mobility, for instance—that changed forever our sense of who we were. It led, most importantly, to a kind of hyper-individualism—the loss of a practical need for our neighbors—now manifested most clearly in the United States. That's become our defining characteristic, as people like Robert Putnam have made clear in recent years. But it dates in some very real way from our ability to burn coal and oil and gas, and the set of decisions we've made about how to do so.

5. With respect to present and future inquiry, how can the most important philosophical problems concerning technology be identified and explored?

Here are two relevant thoughts to consider.

First, look for the things with the possibility to change who we are. There aren't that many—the internet, obviously, and global warming, and genetic engineering, and probably a short list from there. Don't buy in to the standard academic approach exemplified by, say, the bioethics profession—i.e., this new technique is similar to something we've done in the past so it's not much of a discontinuity. Instead, try to imagine how it plays out—what it means for power relationships, for felt human experience. In an interesting way, the best philosophers of technology have been science fiction writers, and it's the reason the scifi section is the most dystopian part of the bookstore. They've had to take human beings – their characters – and hold them up against the scale of new technologies. We should pay attention to the fact that in one account after another the scale of those technologies overwhelms the human scale, obliterates the personal.

Second, the technology we need most badly is the technology of community—the knowledge about how to cooperate to get things done. Our sense of community is in disrepair at least in part because the prosperity that flowed from cheap fossil fuel has allowed us all to become extremely individualized, even hyperindividualized, in ways that, as we only now begin to understand, represent a truly Faustian bargain. We Americans haven't needed our neighbors for anything important, and hence neighborliness—local solidarity—has disappeared. Our problem now is that there is no way forward, at least if we're serious about preventing the worst ecological nightmares, that doesn't involve working together politically to make changes deep enough and rapid enough to matter. A carbon tax would be a very good place to start.

Selected Bibliography

The Age of Missing Information (Random House, reprint, 2006)

The End of Nature (Random House, reprint, 2006)

Wandering Home: A Long Walk Across America's Most Hopeful Landscape:Vermont's Champlain Valley and New York's Adirondacks (Crown, 2005)

Enough: Staying Human in an Engineered Age (Times Books, 2003)

Long Distance: Testing the Limits of Body and Spirit in a Year of Living Strenuously (Plume Books, 2001)

Hundred Dollar Holiday: The Case for a More Joyful Christmas (Simon and Schuster, 1998)

Maybe One: A Case for Smaller Families (Plume Books, 1999)

16

Carl Mitcham

Professor of Liberal Arts and International Studies

Colorado School of Mines
USA

QUESTIONS OF A DIFFERENT ORDER

Being uneasy about placing autobiography in the forefront, I'd like to respond to these five questions out of order — even if, paradoxically, this appears to accent the personal rather than subordinate it. After all, paradox is a fitting introduction to philosophy.

3. What, if any, practical and/or social-political obligations follow from studying technology from a philosophical perspective?

There is no easy answer to this important question, which assumes a distinction between philosophy and practical or political life. Although there are certainly reasonable grounds for such an assumption, it is also a belief that deserves philosophical scrutiny.

Socrates, for instance, saw philosophical questioning as his fundamental political activity and vice versa. Something similar might be said for figures as diverse as Cicero and Alfarabi. With Augustine, Hume, or Kant this would be less so. In fact, with Kant something close to a reversal takes place in his defense of a principled reconfiguration of the practical. So what practical obligations might flow from a philosophical approach to technology will depend importantly on what we think might constitute the philosophical study of technology. Does it mean, with Socrates, questioning the artisans and engineers about their work and self-understandings? Or does it involve, with Kant, doing a critical analysis of the preconditions or presuppositions of technology?

In the modern period philosophy has been prolonged into a number of regionalized fields such as the philosophy of science, of religion, of art, and of history — each bearing on what are taken to be semi-autonomous realms of the cultural system. Philosophy of technology aspires to join the fold. Each regionalization includes a deployment of the basic divisions of philosophy — logic, epistemology, metaphysics, ethics, and political philosophy. Different regionalizations will nevertheless give different weights to these basic divisions. For instance, epistemological issues play a larger role in the philosophy of science (What is scientific knowledge?) than in the philosophy of art. Likewise, ontological issues (Does God exist?) play a larger role in the philosophy of religion than in the philosophy of history.

Among the most prominent issues in the philosophy of technology are those related to ethics and political philosophy. Precisely because the scientific making and using of multi-scale products, processes, and systems defines the milieu in which we now live and act, it is only appropriate that modern technology should become a subject matter for ethics. What is a good technology? What is the right way to undertake technological production and utilization? How can technology be designed to be more just? Indeed, such questions have been increasingly raised within the technologies of engineering, biomedicine, and computers. This is the case especially when such technologies have been associated with potentials for large-scale destruction (nuclear weapons and environmental pollution) or human modification and enhancement (bio- and genetic engineering). In such cases ethical reflection itself appears to be a socio-political obligation.

Is it sufficient to promote simply the ethical reflection process — along with, of course, a commitment to its improvement? Can anything more substantive be said about obligations that might flow from ethical reflection? Philosophical reflection on modern technology — from issues of conceptual analysis and the logical structures of technology, through the theory of technological knowledge and the ontology of artifacts, to those of ethics and politics — points toward three substantive moral imperatives: for increased consciousness, for caution or moderation, and for broadened participation in technological action.

Take a simple example: When walking, we do not have to think too much about it. If I bump into someone, serious injury is unlikely. Riding a bicycle already requires an increased degree of attention because of what an accident might do to me or others — an attention that must be brought to bear, as it were, from outside and imposed by myself on top of my bicycling skill. In driving cars the bar of attention is raised higher still and has to be imposed by law. Flying airplanes requires extensive thinking in preparation for flight on the part of pilots, individual or corporate owners, the engineers who design aircraft and airports, and the government that is called on to regulate many of the interrelated operations. With regard specifically to engineering, this has been formulated as a duty plus *respicere*, to take more into account.

With regard to moderation, surely there are occasions when reason and morality both suggest the need to slow down in order better to consider the options and consequences of a course of technological action. Just as we slow down the car when trying to read a street sign, was it not right to have slowed down the recombinant DNA manipulation of micro-organisms in order to consider the possible ramifications, as was done by molecular biologists during the moratorium of 1974–1975? Are there not other cases in which slowing things down might not be morally appropriate?

Another illustration of this two-fold ethical interest in thinking things out and moderation might be found in the problem of increasing obesity in advanced technological societies. Before the advent of the mass production and distribution of food products, many people were guided toward healthy eating simply by natural availability — and received sufficient exercise in the course of normal working days. Once industrialized agriculture made processed foods available 24/7, consumers have had to undertake a conscious, critical assessment of their eating habits, moderate some of them, and explicitly schedule in physical exercise.

The widely discussed sustainability concept and precautionary principle, as they are brought to bear in technology, entail further arguments for increased consciousness and moderation. Sustainability has become a vision of the good that promotes increased awareness regarding the consequences of actions in order to balance the values of environmental protection and economic development. The precautionary principle is a version of the Hippocratic imperative, "First, do no harm"; in the name of the goodness of that which already exists, this principle seeks to qualify the attractions of change and enticements to technological progress.

With regard to public participation in decision making about the construction of our technological world: The argument is that just as citizens have a right to elect those who will govern them, so do individuals have a right to exercise more supervision than accorded by the free market over the design of technologies that inevitably (if unintentionally) influence the quality and character of their lives. Issues of distributive justice in relation to consumer goods are now complemented by issues concerning just distribution of the risks of those goods and their production. Especially in relation to the just distribution of risks, participatory technological design processes would seem to be obligatory. In the words of Steven Goldman, "No innovation without representation." Robert Nozick's libertarian rejection in *Anarchy, State, and Utopia* (1974) of the general principle of "having a say over what affects you" under appreciates the technological transformation of the notion of private property.

Insofar as such arguments deserve to be taken seriously, they also deserve to be examined philosophically. The challenge for the philosophical assessment of technology is to help clarify and better formulate precisely such general ethical concerns for both consumers and citizens. What if any is the difference between technical and moral requirements for increased consciousness? What particular forms might the vague notion of a duty *plus respicere* really take? How strong is such an imperative in relation with others? Is there a point beyond which attempts to practice such an obligation lack practical wisdom? Could human-computer symbiosis be a proper response to these limits? How is moderation to be justified or operationalized, especially under conditions of competition or crisis? What ethical responsibilities are entailed by participatory decision making? Ought not increased thinking, moderation, and participation be projected as well into professional-technical role responsibilities?

But what is the practical value of the philosophical prosecution of such questions? Will clarifying responses be sufficient to help us lead better lives? In the world of advancing technology, is there not a philosophical challenge to take such issues beyond the level of personal ethics and extend them into the formulation of policy, especially science and technology policy? Ethics alone is not enough.

The insufficiency of ethics deserves to be emphasized as a phenomenon of social-political relevance in any philosophical approach to technology that aspires to provide moral guidance for the engineering of the world. The dialectic here is foreshadowed in the movement in Aristotle from ethics to politics. In many instances we find ourselves lacking policies to alter in any basic manner a trajectory of technological change. As Hans Jonas observed in *The Imperative of Responsibility* (1984), echoing an argument at the center of Plato's *Republic*, it may be that only authoritarian rule by philosophy will be able to redirect or moderate a potentially disastrous technological momentum. Are we not often frustrated more than ennobled by our ethical reflections and blocked in pathways we might chose to reach out for the good? Does this reveal something about the special character of a technoscientific culture — if not some deep resistance to ethical guidance then a fundamental techno-moral commitment of its own? Or has ethics, in dissatisfaction with its own powerlessness, perhaps been infected by the technological mentality? Could the turn to policy be but another manifestation of the tendency of liberal democracy to deprive philosophy of depth and significance?

4. If the history of ideas were to be narrated in such a way as to emphasize technological issues, how would that narrative differ from traditional accounts?

The history of ideas (*Geistesgeschichte* and *histoire de la pensée* being related terms) is a problematic project worthy of philosophical attention in its own right. In what sense does it subsume the history of philosophy, of religion, of science, or of art? According to the U.S. historian Crane Brinton, writing in the *International Encyclopedia of the Social Sciences* (1968), the history of ideas is but one of three types of intellectual history. In the first case, intellectual history just tries to figure out who wrote what and when. In the second, there is an effort to map the interrelations of ideas. It is to this that Arthur O. Lovejoy gave the name history of ideas in the strict sense, as when he traced historical shifts among a multitude of overlapping meanings in the metaphysical concept of a chain of being linking together the elements of the cosmos. In the third, efforts are made to understand the relations between what human beings think or say and how they act.

With regard to technology, the history of ideas could thus seek to map out ideas in technology or ideas about technology. In the former instance, one could imagine, for instance, careful disambiguations of the concept of energy or design within one or more fields of engineering. In the latter, the focus would be more on social or political ideas about machines or engineers. The truth is that histories of ideas in both these senses are alive and well in the field of the history of technology. Yet in both cases the historical analysis will be too limited if it does not go beyond narrating changes in conceptual interactions over time to critical interpretation of those interactions both in and about technology and the societies in which the technologies are found — including attempts to make assessments of such ideas and interactions.

Contributions to philosophy and technology studies from the history of ideas have been limited. "Technology" was not included in Mortimer Adler's list of "Great Ideas." The entry on "Technik" in the German *Wörterbuch der philosophischen Begriffe* (1927–1930) is no more than etymological, and that on "Technology" (by D.S.L. Cardwell) in the original *Dictionary of the History of Ideas* (1968) was simply an 8-page review of the history of some ideas internal to modern technology. The entry on the same term (by Aristotle Tympas) in the *New Dictionary of the History of Ideas* (2005) is a complementary four-page discussion of the ideas about technology associated with arguments for and against technocracy and technological determinism.

One classic example of a philosopher taking a more expansive approach to the history of relations between ideas, lifeworld, and the good is provided by Giambattista Vico, whose work has implications for understanding technology, even though he does not directly discuss technology as such. More recent approaches that might also be taken as instructive can be found in the conceptual archeologies of Michel Foucault and Ivan Illich or the philosophical criticism of ideas by Isaiah Berlin and Donna Haraway. In each case, however, it will be crucial to use the model to further philosophical engagement with technology in ways less than fully present in the original.

Drawing especially on the inspiration of Illich, for example, consider the following notes for a critical history of the idea of tool or *instrumentum*. In Martin Heidegger's most well-known philosophy and technology text, "Die Frage nach der Technik," he begins by noting how technology is a complex activity, object, and volition that is commonly thought together under the combined categories of means and human activity. To this Heidegger gives a Latin term and conceives technology as *instrumentum*.

Heidegger appears both to accept and to oppose this "instrumental or anthropological definition." After affirming its correctness, for instance, he asserts that this "correct instrumental definition" fails to point out the essence of technology. In response, he argues an alternative presentation of technology as a kind of ?λθεια, truth understood as disclosure or revealing. This alternative presentation is developed through a typically Heideggerian maneuver that interrogates instrumentality by turning to a reflection on Aristotle's four causes. For Heidegger, the rethinking of instrumentality in terms of causality appears non-problematic. That is, Heidegger fails to mention how *causa instrumentalis* — as distinct from *causa materialis*, *causa formalis*, *causa finalis*, and *causa efficiens* — has its own history, a history distinct from that of the other causes and one that only came to fruition long after Aristotle.

The Greek equivalent of *instrumentum* is generally taken to be ∠ργανον or *organon*. We should nevertheless be uncomfortable about any easy translation, since the classical Greek word can refer to both an organ of the body as well as an artifact grasped and used as an implement or tool. Indeed, the Greek *organon* is a root of the English "organic" and "organism," not to mention a kind of musical machine.

The closest Aristotle came to putting forth an idea related to what will come to be called the *causa instrumentalis* occurs immediately after he distinguished the four causes of coming to be. Along side these, he writes, are the "intermediary becomings" in which, for instance, "a physician causes health by use of some *erga* or *organa*, therapeutic procedures or drugs" (*Physics* II, 3; 194b35–195a3). These procedures and drugs are not properly subsumed within any of the other four causes, but are intermediaries between the efficient cause of the physician, the final cause of health, the material cause of the flesh, and the formal cause of the soul. Earlier Greek references to intermediary causation occur at a number of points in Plato. In the *Phaedo* (99a ff.) Socrates criticizes the idea of the body as primary cause. Later in the *Statesman* (281d ff.) there is a distinction between αται (causes) and ξυνατια (collaborative causes). Finally, in the *Timaeus*, near the end of a discourse on the works of reason that constitute the coming to be of the human soul and body, the speaker identifies what he calls "secondary causes which the god uses as subservient in order to achieve the best that is possible" (36d).

In this capacity as a means by which the divine relates to the world, *causa instrumentalis* comes increasingly to prominence in Jewish, Islamic, and Christian theology as a distinct species of causation. As Harry Austryn Wolfson analyzes the history of this idea in his great study, *Philo* (1947), instrumental causality becomes a way to preserve God as both fully transcendent and creator of the world. God works through secondary or instrumental causes. He only touches the creation with the proverbial ten foot pole of angels and other intermediaries. This Hellenistic Jewish idea of secondary or instrumental causation is picked up and elaborated in the Islamic theology of al-Kindi (9th century CE), Ibn Sina (or Avicenna, late 10th and early 11th centuries CE), and Ibn Rushd (or Averroes, 12th century CE). In each of these instances the aim is to defend the absolute character of divine power and allow for the complexities of physical and biological causation.

In scholastic Christian theology, instrumental causality is prominent in at least two different kinds of relations: the relation between ministers and the sacraments they perform, and the relation between angels and the heavenly spheres of the planets they move. For present purposes, consider only the case of the sacraments. In the *Summa theologiae* (III, question 62, article 1), Thomas Aquinas asks in what way the sacraments cause grace. According to the common objections, the sacraments are not a cause but simply a sign of grace. According to Thomas, however, following the teaching of Augustine, when baptismal water touches the body it brings about a cleansing of the heart. And since the heart is not cleansed except by grace, the sacrament of baptism (and *eo ipso* the other sacraments as well) must cause grace.

To explain how this is possible Thomas develops the notion of the *causa instrumentalis* as a special kind of *causa efficiens. Causa efficiens* or *causa agens* can be either what he calls "principal" or "instrumental." "The principal cause produces its effect in virtue of its form, to which that effect is assimilated, as fire warms in virtue of its own heat." The instrumental cause, by contrast, acts not in virtue of its own form, but solely in virtue of the motion by which the principal agent moves it. Hence the effect has a likeness not to the instrument, but rather to that principal agent: as a bed does not resemble the axe which carves it but rather the art in the mind of the artificer.

The instrumental cause is, as it were, neutral. It takes on the intentions and actions of its prime user who is, it is crucial to note, not the minister or priest but God who uses them and the priest to administer the sacraments.

This is a remarkable argument; one that it may be suggested foreshadows a uniquely modern notion of technology. One implication is that the ministers of the Church can validly confer the sacraments independently of their own moral state or character, thus radically dis-embedding the sacramental activity from the spiritual lifeworld. Does this theological idea of the tool as a neutral instrument through which a transcendent God acts into the world not foreshadow a modern faith in technologies as neutral instruments through which will flow without distortion the intentions imparted to them by their human makers and users? When Edward Teller proposed using hydrogen bombs to do geological engineering was he not only thinking of himself as god-like but also the nuclear devices as pure means that would not in any way introduce a reality of their own into the harbors and canals he wished to create?

The notion of disembedding here draws on that of the economic historian Karl Polanyi, who describes traditional economic activities as deeply embedded in a contextualizing, cultural manifold. The modern economy is distinguished precisely by being disembedded from religious customs, political and social orders, aesthetic aspirations, and other aspects of culture. In a like manner, the scholastic Christian theology of the causal character of the sacraments disembeds them from a larger socio-moral nexus and turns the sacraments into what will later be called a resource. It is against the background of this matrix of ideas that there emerges the idea of technology as neutral instrumental cause or tool. Foreshadowing the way nature is turned into a resource by modern science and technology, and the ways men and women have become human resources for corporations and governments, the sacraments were given the status of spiritual resources within the Christian community. What for Heidegger is the distinctive truth of modern technology — its making available of energy and materials for human manipulation — is adumbrated in the scholastic making available of grace. In the summary judgment of Illich, *Corruptio optimi quae est pessima*, the corruption of the best is the worst.

What also follows from such a briefly sketched and quite partial history of the idea of instrumentality is that despite the overwhelming and dominant contemporary assumption, tools, tool-using and making are not ahistorical features of being human. As the cultural critic Lewis Mumford has maintained, in an argument deepened by the social anthropologist Tim Ingold, the idea of tool using as coeval with humanity depends on an anthropological anachronism that would read back into the prehistorical record our own experiential fascination and ways of relating to technology.

Nevertheless, the danger of the history of ideas pursued in such a manner is that it can easily become little more than a parasitic scholasticism wallowing in the detritus of words and usages. Only with proper cultivation can it help liberate the mind from the limitations of any particular use and ascend to a richer understanding of meanings and their potential for transformation.

5. With respect to present and future inquiry, how can the most important problems concerning technology be identified and explored?

It is by means of interdisciplinary engagements among historians, scientists, engineers, philosophers, and social scientists — not to mention physicians and lawyers — that many philosophical problems or issues concerning technology have been identified. These issues include the need for conceptual clarifications among different types of technology, relations between science and technology, the theory of technological (including medical and engineering) knowledge, the ontology of artifacts, the mutual interactions between technology and culture (including religion, politics, work, art, entertainment), the anthropological foundations and historico-philosophical origins of technology, and questions focused on goodness, obligation, justice, and the beautiful in and about technology.

The exploration of these issues nevertheless remains to be pursued and deepened especially in relation to on-going technological or technologically influenced change in multiple dimensions and at various scales: economic, environmental, global, genetic, and nanomaterial. That philosophy and technology studies has not precipitated out into a strong academic discipline after the manner of the philosophy of science, of religion, or of art may reflect how much there is to do. What has happened instead is the emergence of what might be called regionalizations of the philosophy of technology: environmental philosophy, philosophy of medicine, philosophy of computers, etc. In all instances what will continue to be crucial is interdisciplinary, multidisciplinary, and transdisciplinary work — which is why the theory of interdisciplinary, too, has a role to play in philosophy.

One way to pursue and deepen philosophy and technology studies would be through an interdisciplinary history of ideas involving what might be called an existential comparison and contrast. One of the under-considered issues is that of suffering. Consider, for example, a possible comparative assessment of the existential implications of Christian and Buddhist world views of suffering in relation to modern technology.

It is extremely difficult to compare the moral orders of the Christian versus the Buddhist cosmos. In the Christian view, at least as it has developed in the modern West, human suffering is seen as a phenomenon not to be accepted but to be heroically struggled against. There is something horrible and unacceptable, for instance, about letting an ill-formed infant or sick child die, about not donating an organ to someone in need, about not donating one's body to science so that death can be transmuted into a possible means for the cure of illness or disease. This is especially true insofar as those who might accept such events for others would not themselves forego medical interventions or live with their own suffering and death. In the words of Paul Farmer, founder of Partners in Health (PIH) and a humanitarian health activist, this is an "area of moral clarity":

"When a person in Peru, or Siberia, or rural Haiti falls ill, PIH uses all of the means at its disposal to make them well — from pressuring drug manufacturers, to lobbying policy makers, to providing medical care and social services. Whatever it takes. Just as we would do if a member of our own families — or we ourselves — were ill" (Partners in Health, Vision Statement).

From the Buddhist perspective, by contrast, the struggle against suffering takes a much more interior form. Suffering is viewed as a result of internal cravings more than external circumstances; internal desires are to be assessed and transformed more than the world in which they seek satisfaction. Indeed, to the Buddhist there are even respects in which there could appear to be something horrible and unacceptable about trying to keep an ill-formed infant or sick child alive by all available means, about donating one's organs or body to science out of a determined commitment to do good. The human desire for life is to be acknowledged, but we must struggle to place it in proper perspective, to avoid excessive attachment, even while experiencing and expressing compassion for life and its attachments. The Buddhist world is suffused by what might be described as a kind of compassionate melancholy. Buddhist compassion, in contrast to Christian charity, does not entail the relief of suffering so much as the cultivation of an insight into its foundations that allows for a stepping aside from both pleasures and sufferings.

The anthropologist Michael Carrithers, in an ethnographic study of the way of life of *The Forest Monks of Sri Lanka* (1983), offers one entre into a perspective that is deeply foreign to European world views. At one point Carrithers describes how the traditional Jatake tales of the many lives of the Buddha weave together "the other-worldly values of asceticism and the this-worldly values of the family" (p. 90) so as to combine a serious compassion for the family and the suffering experienced when one of its members renounces the world, with the presentation of such renunciation as a heroic act that the family accepts through its own painful but heroic affirmation.

As Carrithers summarizes the situation:

"The struggle for perfection by the Bodhisattva in the Jataka cycle is cast so that the dearest and most common desires of the flesh and family life are called upon to bear witness to the heroism of sacrifice and renunciation. If your own wife and child are dear to you, say the Jatakas, how much more precious must it be to renounce them? And if your own flesh is dear to you, how much more precious must it be to sacrifice it? Hence, through the countless births, the Bodhisattva sacrificed a mountain of flesh, an ocean of blood, as the Jataka poets are fond of saying; while the woman who was his wife in so many births, and whom he left in his last birth to become the Buddha, wept an ocean of tears for him. The [hearer of such tales], of whatever condition or sex, can therefore identify with underlying Buddhist morality because the emotional charge inherent in worldly life animates the (more noble) values of renunciation. ... It is quite possible ... to use the supreme status accorded the Buddha and his wilful renunciation to cast a glorious, if melancholy, light on family life itself..." (p. 92).

In his interpretation of the Jataka tales, Carrithers emphasizes how renunciation is presented in a doubly erotic context: the love of the mother for her son and the eros of renunciation itself, and "how asceticism and renunciation — the taking of special vows, the voluntary submission to pain, the leaving of the pleasures of the world — appear over and over again" (p. 94). The trick is to accept actively and not to become passive in the acceptance of evil. One must struggle within limits to save dying children, and genuinely suffer the loss of children on those occasions when they die. One must not replace a compassion within limits with a dispassionate or unfeeling acceptance of death.

Despite the apparent foreignness of this Buddhist sensibility to the Christian world view, one can find traces of it in the Christian tradition, especially the early stories of Christian martyrs. One example would be the story of Saint Perpetua. In the 2nd century Perpetua, a young mother, who was condemned to death because of her faith, was implored by her father, for his sake and the sake of others, to foreswear the path of martyrdom. "Look upon your brothers; look upon your mother and mother's sister; look upon your son, who [because he still nursed at her breast] will not survive after you." The response, contained in the *Vibia Perpetua* that she wrote in prison, has a zen-like quality: "Do you see this pitcher? Can it be called anything other than what it is? No, he answered. In the same way I cannot call myself other than that which I am, a Christian."

1. Why were you initially drawn to philosophical issues concerning technology?

Looking back on my formative years, an observer might well describe a natural temperament in tension between puritanical idealism and aspirations for understanding of a world in which I did not feel at home. That which was my own, which for Aristotle is the first name of good, was experienced as strange if not foreign. From the beginning I was unsure of who or where I was, especially in relation to the technologies that increasingly surrounded me. Paul Goodman's *Growing Up Absurd* (1960) offered a sociological analysis of something that felt more than sociological.

My memory includes a number of relevant stories, one of which goes as follows. I grew up in Dallas, where my father was a mechanical engineer, but I spent formative periods of my childhood on my uncle's ranch in Fredericksburg, Texas. For my uncle a typical day began with going to the barn, getting on the tractor, and spending much of the day plowing or harvesting a field. For my father the day involved getting on a bus and returning in the evening. Both my uncle and my father were said to be "working." But it was unclear to me how the two were doing anything like the same thing.

One day, in response to a question about what he did at work, my father took me to the Otis Elevator Company, his place of employment. I watched him make drawings on a drafting table, answer the telephone, read and write memos, attend meetings, and talk to people. He was a mechanical engineer who supervised the installation of elevators. But then I discovered that my uncle actually had a degree in engineering too, in civil engineering, and that before he married and went with my aunt to live on part of her family's ranch, he had worked for the Texas Highway Department in road construction. My uncle once even described his work now, on the farm, as that of an agricultural engineer.

How could all these different activities, these different examples, be work? What was work itself? What was engineering such that mechanical and civil and agricultural engineering could all be types of engineering?

Then in high school I read Mahatma Gandhi's autobiography, *The Story of My Experiments with Truth* (first published in 1927–1929), and Jawaharial Nehru's autobiography, *Toward Freedom* (from 1936). Gandhi criticized modern industrialism and engineering as destructive of human well being, and Nehru praised it as the hope for future well being. Somehow I sensed that in my own life there was a Gandhi truth (the way of life on my uncle's ranch was going to disappear) and a Nehru truth (we were becoming more and more powerful and wealthy as a result of technology). But insofar as these were competing truths, which was to be given priority?

My first year at the university I was assigned to read Alfred North Whitehead's *Science and the Modern World* (1925) where, in the first chapter, he confidently asserts: "More and more it is becoming evident and what the West can most readily give to the East is its science and its scientific outlook." Against the background of the debate between Gandhi and Nehru, I doubted this was the case, and noted especially the short shrift given, especially in the last chapter on "Requisites for Social Progress," to the problems of technology. Perhaps there was a correlate in the realm of practice to what he identified in the realm of theory as the "fallacy of misplaced concreteness."

In 1962, the same year as the Cuban missile crisis and the publication of Rachel Carson's *Silent Spring*, I was at Stanford University taking Donald Davidson's course on the philosophy of action. In an encounter, whose real and imaginary aspects I am no longer able fully to distinguish, I asked Davidson why, in a course on action, we were only dealing with issues such as what is the difference between my arm going up and me raising my arm (Ludwig Wittgenstein, *Philosophical Investigations*, I §621)? Why were we not asking questions about pollution and nuclear weapons? His response: "Carl, those are not *philosophically* interesting questions."

Two years later when I stumbled into classes taught by former students of Leo Strauss and Etienne Gilson on Plato and Thomas Aquinas, respectively, I realized that there were other paths in philosophy. What became philosophically crucial was finding a way to reach out to the good as known by the ancients in the midst of the modern technological world.

The cumulative influence of such experiences drew me little by little to something I persist in calling philosophy and technology studies, out of fear that "philosophy of technology" will become a professionalized specialization turned in on itself, one that will fail to address the broad range of issues I experienced as calling for critical philosophical reflection.

2. What does your work reveal about technology that other academics, citizens, or engineers typically fail to appreciate?

The first phase of my work in philosophy and technology studies rested on two discoveries: (1) that scientists, engineers, and social scientists were often more interested in the kinds of questions I wanted to examine than were professional philosophers, and (2) that there nevertheless existed a tradition of philosophical reflection on technology in Europe that was underappreciated in the United States. Bibliographical and editing work during this period sought to promote greater recognition of the second, while my scholarly writing made an effort to use what had been learned from European philosophers as diverse as Friederich Dessauer, José Ortega y Gasset, Heidegger, and Jacques Ellul to open a more substantive dialogue with the technical community. In both instances interdisciplinary connections were crucial — with science, engineering, history, literature, religion, and art.

A second phase of scholarship turned toward questions of ethics and politics. Here my effort has been to argue the need for synthesis between the diverse regionalizations in philosophy and technology. There are not only technologies in the plural; there is technology in the singular that exists in different forms throughout nuclear, biomedical, computer, genetic, and other technologies.

Does this work reveal anything about technology that other academics, citizens, or engineers typically fail to appreciate? I doubt it. What is important for me is what is surely important for everyone who is not ensnared in technology as a good in itself: It is to try to think technology, to take what is fundamentally a kind of making and using of artifacts, and seek to understand it. This understanding is oriented not so much toward improving or advancing technology as simply understanding it.

Biographical Note on Institutional Affiliations

Mitcham earned degrees from the University of Colorado (BA and MA) and Fordham University (PhD). He has taught at Berea

College (Kentucky), St. Catharine College (Kentucky), Brooklyn Polytechnic University, Pennsylvania State University, and the Colorado School of Mines, where he is currently Professor of Liberal Arts and International Studies. Additionally, he is on the adjunct faculty of the European Graduate School and an associate of the Center for Science and Technology Policy Research at the University of Colorado, Boulder. Visiting positions include the University of Puerto Rico, Mayagüez; the Universities of Oviedo and of the Basque Country in Spain; the Universities of Tilburg, Twente, and Delft in the Netherlands; and the Consortium for Science, Policy, and Outcomes at Arizona State University.

Bibliography

I often have the sense of publishing too much—that it would have been better to publish less of higher quality, and that in too many instances I have devoted time to editing which would have been better spent thinking, teaching, and writing. At the same time, much of the editing was undertaken as a service to the philosophy and technology studies community. What follows is a selective list of books only.

Philosophy and Technology: Readings in the Philosophical Problems of Technology. New York: Free Press, 1972. Pp. ix, 399. (Co-edited with Robert Mackey.) Paperback reprint, with updated Select Bibliography, New York: Free Press, 1983. Pp. xii, 403.

Bibliography of the Philosophy of Technology. Chicago: University of Chicago Press, 1973. Pp. xvii, 205. (With Robert Mackey.) Reprint with index, Ann Arbor, MI:Books on Demand, 1985.

Theology and Technology: Essays in Christian Analysis and Exegesis. Lanham, MD: University Press of America, 1984. Pp. ix, 523. (Co-edited with Jim Grote.)

Philosophy and Technology II: Information Technology and Computers in Theory and Practice. Boston: D. Reidel, 1986. Pp. xxii, 352. (Co-edited with Alois Huning.)

¿Qué es la filosofía de la tecnología? Barcelona: Anthropos, 1989. Pp. 214. Russian version: *Chto takoe filosofiia tekhniki?* (Moscow: Aspekt Press, 1995). Chinese version, with a special introduction: *Jishu zhexue gailun* (Tianjin, China: Tianjin Science and Technology Publishing House, 1999).

Ethical Issues Associated with Scientific and Technological Research for the Military. Proceedings from a January 26-28, 1989, conference. Annals of the New York Academy of Sciences, vol. 577 (December 29, 1989). Pp. xvi, 277. (Co-edited with Philip Siekevitz.)

Philosophy of Technology in Spanish Speaking Countries. Philosophy and Technology, vol. 10. Boston: Kluwer, 1993. Pp. xxxvi, 318.

Thinking through Technology: The Path between Engineering and Philosophy. Chicago: University of Chicago Press, 1994. Pp. xi, 397.

Engineer's Toolkit: Engineering Ethics. Upper Saddle River, NJ: Prentice Hall, 2000. Pp. x, 131. (With R. Shannon Duval.)

Visions of STS: Counterpoints in Science, Technology, and Society Studies. Albany, NY: State University of New York Press, 2001. Pp. vi, 170. (Co-edited with Stephen H. Cutcliffe.)

La ética en la profesión de ingeniero: Ingeniería y ciudadanía, Santiago, Chile: Universidad de Chile, 2001. Pp. 161. (With Marcos García de la Huerta.)

The Challenges of Ivan Illich: A Collective Reflection. Albany, NY: State University of New York Press, 2002. Pp. viii, 256. (Co-edited with Lee Hoinacki.)

Toward a Philosophy of Science Policy: Approaches and Issues. Philosophy Today, vol. 48, no. 5, Supplement 2004. Pp. 124. (Co-edited with Robert Frodeman.)

Encyclopedia of Science, Technology, and Ethics. 4 vols. Detroit: Macmillan Reference, 2005. Pp.cxiv, 2378. (Editor-in-chief.)

Two works in progress:

Technology and Religion: Oppositions, Sympathies, Transformations. Westport, CT: Greenwood Press, forthcoming.

Science, Technology, and Ethics: An Introduction. Cambridge, UK: Cambridge University Press, forthcoming.

17

Andrew Pickering

Professor, Department of Sociology

University of Illinois
USA

---

1. Why were you initially drawn to philosophical issues concerning technology?

I came to technology by an indirect route. In the late 1970s, my earliest work in STS was in science studies proper, looking in different ways at my former specialty, elementary particle physics. I tried my hand at what was then the favoured methodological genre, the controversy study, and I was surprised at what I found. The standard rhetoric was that in controversies nature factors out. All parties look at the same 'nature,' so divergent accounts of nature have to be understood in terms of sociological rather than material variables. That turned out not to be true in the cases I looked at. None of the classical sociological stories seemed to have much purchase on my studies, nor, in fact, did it make sense to see different scientists as confronted by the 'same' nature. Instead I found scientists giving very different accounts of the world using very different instruments, and I further concluded that specific accounts hinged very much on the details of struggles with apparatus to get it to perform in a way that made sense (Pickering 1981).

At that time none of this was easy to articulate: the Strong Programme in the sociology of scientific knowledge, for example, echoed the dominant discourses in history and philosophy of science in its focus on scientific theory, representations and ideas. I felt that we had to do more to inject the material culture of science—instruments and machines, and their performance and powers—into the discourse of STS. I did my best in my first book, *Constructing Quarks* (1984), by including, for example, a chapter on 'Producing a World,' which showed how different regimes of instruments and data-processing algorithms sustained different general pictures of the world of elementary particles—the 'old physics' and the 'new physics' which replaced it in the early 1970s.

An important next step for me was a 1987 essay called 'Living in the Material World.' There I went back to an earlier study to analyse the production of knowledge in quark-search experiments in terms of an open-ended back-and-forth performative *dance of agency*, as I later called it, between the experimenter and his apparatus, in a process that I came to call the mangle of practice. 'Material World' developed over time into another book, *The Mangle of Practice; Time, Agency, and Science* (1995) by which time I had realised that science and technology looked much the same at the material level—machines and instruments are integral to both—a point that I sought to make by offering the same form of analysis of the introduction of computer-controlled machine tools in the workplace as I did of episodes from the history of science and mathematics. And that was the point at which I began to feel that technology was perhaps even more interesting than science from my point of view, because it thematises manoeuvres in the field of human and nonhuman agency without too many of the distractions inherent in scientific knowledge production.

After *The Mangle* I lost interest in doing science studies in the classical mode, though I worked through two case studies of large-scale shifts in science, technology and society—of the coupling of academic chemistry and the synthetic dye industry in the later 19th century, and of science and the military in and after World War II—arguing that these episodes can be seen as instances of a decentred and posthumanist *macro-mangling* of the material, the social and the conceptual (Pickering 1995b, 2005). My current interest in the history of cybernetics grew out of the latter project, and I am now finishing a book on cybernetics in Britain as elaborating a mangle-ish ontology in brain science, psychiatry, engineering, robotics, biological computing, the arts, entertainment and spirituality (the list goes on) (Pickering forthcoming a; see Pickering 2002 for an overview of this project).

So that's the answer: I slid into the philosophy of technology from science studies and an interest in the material performances of things, and my interest in technology still shades into many other realms, including science.

2. What does your work reveal about technology that other academics. citizens or engineers typically fail to appreciate?

That is a dangerous question to answer, expecially as it concerns my colleagues in science and technology studies. I don't want to claim any dramatically unique insights. But I could make a couple of remarks. One is that the modern academic disciplines, like modern commonsense, tend to make a clean split between people and things—the social sciences are only about people; the natural sciences and engineering are only about things on their own—and this dualism makes it hard to grasp any reciprocal coupling of technology and society. On the other hand, the social world has been changing rapidly since the Industrial Revolution and it would be very surprising if that did not have a lot to do with changes in the field of technology, and vice versa. The great strength of the kind of analysis I have developed in and since *The Mangle* is that it is non-dualist—it speaks to the co-evolution of machines and social structures—and it thus helps us to understand who we are now and how we got to be this way. That said, I am not the only person in the field of STS to have developed such an analysis, and I should acknowledge my own debt especially to the pioneers of actor-network theory: Michel Callon, Bruno Latour, John Law. It was a key event for me when I first tried to go through Latour's *Science in Action* (1987) in my teaching. Likewise, although it took me a long time to see the continuities between science and technology that I just mentioned, something like this has been implicit in the actor-network approach from the start.

That isn't to say that my work is the same as other people's, of course. One significant divergence might concern how to think about the coupling of people and things. My own way is to think about material agency—the lively and unpredictable performances of machines and nature itself—as that with which human agency intersects and sometimes collides in transformative encounters. I find this the most illuminating way to proceed, but others don't. In traditional sociology, 'agency' has somehow become equated with goal-oriented action, so my way of speaking is simply verboten. But the actor-network theorists are also nervous about this topic, too, for reasons I don't fully understand but which might have to do with a certain nervousness concerning realism and scientific authority. In the famous 'chicken debate,' Harry Collins and Steven Yearley (1992) took Callon and Latour to task for presuming to speak about the powers and capacities of scallops and door-closers: if anyone knows about the agency of scallops it must be scallop-scientists (marine biologists?), so to talk this way must, in the end, mean simply accepting the scientists' own accounts of what the world is like. Callon and Latour (1992) replied that their 'actants' are semiotic and textual rather than material ones, thus avoiding Collins and Yearley's accusation but at the same time, of course, losing their grip on their material referent, which is a shame. I think it is important to keep the materiality of science and technology clearly in focus, and in *The Mangle* (ch 1) I suggested a different response to Collins and Yearley. One can analyse the dance of human and nonhuman agency as a dialectic of resistance to and accommodation of human initiatives, where 'resistance' refers to finite difficulties arising in the course of practice, and we can certainly describe such resistances without repeating scientists' accounts of their origins *or* the resort to semiotics.

Within the overall field of non-dualist STS, then, my approach strikes me as more down to earth and ontological than others, and less focussed on signs, representations and texts, though I don't want to make a big fuss about that. On the other hand, something might be at stake here—but I'll save that for the next question.

3. What, if any, practical and/or social-political obligations follow from studying technology from a philosophical perspective?

This question interests me a lot, having been pressed on it concerning my own work—what follows politically from the mangle? I used to think that the answer was more or less nothing. The mangle is simply an analysis of what happens in the world, so how could it make a difference to feed that story back into the world? But now I think the answer is more interesting and that something important might be emerging here. It's rather a long story, which goes as follows.

My work in STS began with studies of physics, the paradigmatic science of Modernity. And sometime after finishing *The Mangle* I realised that there was a disjuncture between the ontological vision that I had arrived at and the vision that more or less defines the science I was studying. My ontology was a lively, quasi-organic one of interaction and emergence, while physics itself speaks of a cool, stable, repetitive world of fixed entities (quarks or whatever). And this struck me as odd. Since my ontology grew out of my studies of physics, I had to see physics as somehow denying or veiling its own ontological condition—obvious facts about itself. Given its impressive history since the Scientific Revolution this couldn't mean that physics was all a mistake, so I came to see physics—and all the other Modern sciences, and Modern approaches to technology and engineering—as a productive though rather peculiar *stance* in the world—a way of reading the world *against the grain*, so to speak.

This realisation encouraged me to wonder, in turn, about the possibility of a different stance in the world. What would practice that was reflexive to the mangle—that took for granted a quasi-organic ontology instead of a physics-like world of constancy—look like? And another realisation followed: there are many non-Modern projects and traditions in the world—in art, science, engineering, philosophy, spirituality—that indeed act out a mangle-ish rather than a Modern stance (for a quick catalogue, see Pickering forthcoming b). I had, in fact, been interested in many of these, running from Buddhism to complexity theory, for a long time—I had simply not connected that interest before to my work in STS. And this gets us back to my answer to the first question. My current work on the history of cybernetics is driven by a curiosity to see what the mangle-in-action looks like in real-world practice, with cybernetics as an excellent example of how that goes in many fields, including technology and science, but going far beyond that.

We still haven't got to politics, but to make the leap I can now invoke Martin Heidegger. In 1984 I read his famous essay, 'The Question Concerning Technology' (1976 [1954]), for the first time and I could make nothing of it. The second time I read it—perhaps five years ago, in a graduate seminar—I found it immensely suggestive. Heidegger understood Modernity as characterised by the technological project of *enframing*—using machines to turn the world into 'standing reserve' for our human plans and ambitions: the electrical power station straddling the Rhine, reducing this totemic geography to a source of power for factories. Heidegger also understood Modern science as integral to projects of enframing—as setting nature up for domination via the sort of knowledge that goes with the Modern ontology. All of this struck me as historically correct in broad outline.

Finally, Heidegger understood enframing as an enormous danger, from which 'only a god can save us' (Heidegger 1981 [1966]). Two questions strike me about this. The first concerns the nature of the danger. I think Heidegger understood it in terms of damage to our very Being, but one can surely spell this out more concretely. *Catastrophe and disastrous failure* seem to me to be the corollaries to enframing, the dark side of Modern science and engineering. I have written about the history of engineering attempts to control the flow of the Mississippi River as a beautiful exemplification of a dance of agency between the river and the US Army Corps of Engineers itself organised around a telos of enframing the river. But in 2005, in Hurricane Katrina, human attempts to dominate the river failed, amidst massive death and destruction in the flooding of New Orleans. The horrendous chaos of contemporary Iraq can likewise be understood as the corollary of a distinctly Modern impulse to dominate other people, to impose a preconceived order ('freedom and democracy') upon them. (This is an ontological point, not a point about specific political motivations.)

On the other hand, I wonder if Heidegger's pessimism was entirely warranted. Do we really have to wait for a god (or poets and artists) to save us? Heidegger had a monolithic understanding of science and technology as necessarily part of the apparatus of enframing, but perhaps we could think differently. There have always been nonModern approaches to science and engineering lurking in the margins of Modernity, and perhaps now would be a good time to take them more seriously. Stafford Beer, for example, the founder of management cybernetics, defined cybernetics as the science of 'exceedingly complex systems,' meaning both systems that are in practice so complicated that we can never hope to dominate them through knowledge, and systems that ineluctably become, so that our knowledge of them is continually out of date.

And here I would emphasise the following. This cybernetic ontology which imagines a world populated by exceedingly complex systems immediately implies a symmetric *respect* for the other (in nature or society): if one cannot dominate the other, the best one can possibly do is pay attention to the other's performances and explore ways of *adapting* to them. (This happens to be Donna Haraway's sense of 'love' in her 2003 book on relations between humans and dogs.) We could call this a stance that thematises *revealing* rather than enframing, in Heidegger's terms—an openness to what the world has to offer us, rather than a determination to bend it to our will. And then two properties of this stance are worth making explicit. First, that *ontology makes a difference*: projects of revealing are very different from enframing projects—and, again, my interest in cybernetics is precisely to explore such differences in a variety of fields in some detail (see, for example, the strikingly imaginative cybernetic approaches to biological computing discussed in Pickering forthcoming b). And second, one aspect of this difference is a certain robustness. In our dealings with the environment, for example, there is an approach called adaptive environmental management which centres on an experimental attitude to *finding out* what rivers and ecosystems want to do, and the potential for catastrophe and disaster would clearly be reduced if this were our paradigm rather than the command-and-control approach of the Army Corps of Engineers' (Asplen forthcoming).

If we understand politics as thinking about the good life, then, this is my answer to the question. I do not think we need to do away with Modern science and engineering, but I would like to challenge their hegemony. Simply pointing to the existence and viability of all sorts of nonModern traditions in science, engineering and elsewhere is one contribution to that, and I would like to think of this as an intervention into their dynamics, a small contribution to fostering their future growth—perhaps by encouraging people to see them along the lines I just sketched out. The more these traditions and their distinctive products emerge from the shadows, the less likely we are to engage in unreflective and potentially disastrous enframing projects, and the more likely we become to engage in projects of revealing in which genuine novelty can actually emerge.

From the perspective of everyday politics, these remarks are about a kind of *sub-politics*: they imagine a world populated by nonModern projects and their distinctive artefacts – adaptive buildings, for example, that respond to their environments instead of riding rough-shod across them – in which the Modern impulse to dominate would seem less and less natural. Closer to the procedural level of what we usually call politics, I could mention that in the history of cybernetics Stafford Beer had several interesting notions about 'geometries' of decision-making that certainly offer a closer approach to genuine democracy than our usual political arrangements (Pickering 2004).

This last point gets me back to the previous question about differences within STS. I have long been fascinated by Bruno Latour's explorations of 'political ecology' (Latour 1993, 2004), and I think the world would be an interestingly better place if reconfigured along the lines he suggests, bringing scientists and politicians together in a single room in decision-making, for example, instead of the present confused oscillation in authority between parliaments and labs. On the other hand, I am struck by the rather humanist quality of Latour's political reading of actor-network theory. In my terms, he seems happy enough with the Modern stance in science and engineering—'We have simply to ratify what we have always done' (Latour 1993, 144)—and his argument is that we just need to *think* about it differently, recognising the decentred quality of science, especially, and accommodating our political processes to that.

In the end, I wonder how far that can get us. As long as we surround ourselves with the products of Modernity—a spatially and temporally linearised and Cartesian built environment populated by non-adaptive machines and artworks—are we ever going to be able to grasp that we live in a world of exceedingly complex systems immune, in the end, to our control? My suspicion is that something more drastic is required to challenge the hegemony of Modernity and its dialectic of domination and disaster—we'll have to *do things* differently and *make other sorts of things* that will remind us of our ontological condition—thinking differently is just not enough. These ideas have no place in Latour's political ecology, and perhaps this is a reflection of the divergence I mentioned earlier, concerning whether we should think of posthumanist decentring in terms of an ontological notion of a dance of agency or by an appeal to semiotics.

4. If the history of ideas were to be narrated in such a way as to emphasise technological issues, how would that narrative differ from traditional accounts?

At last the questions get easier. As a genre, the history of ideas implicitly or explicitly suggests that ideas are autonomous from the mundane world and that they have their own inner dynamics. In reference to science, one has the further suggestion that scientific theory comes first, then materialises itself as technology, which in turn transforms society. This whole complex of ideas is a conceit of Modernity and the Enlightenment. Empirically it is false. Ideas are not at the centre of the action in human history; they have never called all the shots. My own work in the history of science and technology led me to propose that a good way to grasp human history would be to write it as a *history of agency* – a story of the micro- and macro-mangling of the material, the social and the conceptual centred on enduring zones of encounter between human and nonhuman agency, places like the factory and the battlefield (Pickering 1995a, ch 7, 1997). Besides the advantage of empirical adequacy, such an approach would serve to set science in its place. Instead of being the prime mover of history, we could see it as a finite detour (Latour's word) emerging from and returning to the everyday world of production and consumption (Pickering 1995, 2005). And, of course, grasping history in this way points to the sort of nonModern ontology I just discussed as important in breaking the spell of Modernity, theoretically and politically.

5. With respect to present and future enquiry, how can the most important philosophical problems concerning technology be identified?

I can't think of any general answer to that question and probably there isn't one, but here's a line of thought that interests me and that carries on the ontological concern with the world and the self.

If Heidegger thought that the danger of technology lay in sucking us into an enframing relation to nature, one might argue that at the individual level the danger is more one of evasion. The prototypical technology I have in mind here is something like the Sony Walkman, a device that made it possible to replace one's worldly sonic environment with an artificially crafted one. Instead of hearing the world in its indefinite though often tedious richness as you pass through it, you just hear your preselected favourite songs. This sort of technology is proliferating, at least amongst people who can afford it: video games, iPods, 'home entertainment centres,' cellphones (from a different angle, airbags in cars). People (not me, actually) are more and more enmeshed in designer virtual realities and less and less exposed to the hassles and novelties of the world-as-found (and the global economy depends more and more on this state of affairs). We could think of Marx's 'opium of the masses,' except that the dope now cuts us off not just from political reality but from any kind of reality at all except that offered by the 'content providers.'

Why should we worry about this? It seems to me that there is an evident impoverishment of the life-world happening here (cf Hayles 1999 on disembodiment), but if people want to live in a universe of movies-on-demand why shouldn't they? Isn't this what free-market capitalism is all about?—the market that provides just what people want. I have no definitive answer to that question either, except perhaps the following *reductio ad abdsurdam.*

I recently came across a 1950s publication that I had thought to be some sort of urban legend amongst social scientists. James Olds (1958) reviewed experiments on rats with electrodes implanted in their brains; when the rats pushed a pedal their brains received a small electrical jolt. It turned out that when the electrodes stimulated the 'pleasure centres'—tautologously defined by the induced effects—the rats would just push the pedals as frequently as mechanically possible, for hours on end, until they fell asleep from exhaustion. When they woke up they would do the same again.

We could see these rats as choosing to inhabit their own ultimately stripped-down version of virtual reality—terminal bliss—and I, at least, can see this as the telos of the technological developments just mentioned, from the Walkman to the present and beyond. Forget the science-fiction fantasy of having virtual sex with beautiful actresses; I think we all really could get electrodes implanted in our pleasure centres tomorrow (and that I am giving away an idea that could make me fabulously rich).

Thus the reduction to absurdity—this spectacle of human life degenerating into a series of technologically administered electrical shocks to the brain. It revolts me, of course, as it would have Heidegger, but that doesn't settle anything. Perhaps we are discovering that terminal bliss is the telos not just of consumer technology but of the human race as well. The challenge to philosophy might be to reassure us that this can't possibly be the case. I wish I could see how the argument would go . . .

References

Asplen, L. (forthcoming) 'Acting in an Open-Ended World: Nature, Culture, and Becoming in Environmental Management,' to appear in A. Pickering and K. Guzik (eds), *The Mangle in Practice: Science, Society and Becoming* (Durham, NC: Duke University Press).

Callon, M. and B. Latour (1992) 'Don't Throw the Baby Out with the Bath School! A Reply to Collins and Yearley,' in Pickering (1992), pp. 343–68.

Collins, H. M. and S. Yearley (1992) 'Epistemological Chicken,' in Pickering (ed.), *Science as Practice and Culture* (Chicago: University of Chicago Press), pp. 301–26.

Haraway, D. (2003) *The Companion Species Manifesto: Dogs, People, and Significant Otherness* (Chicago: Prickly Paradigm Press).

Hayles, N. K. (1999) *How We Became Posthuman: Virtual Bodies in Cybernetics, Literature, and Informatics* (Chicago: University of Chicago Press).

Heidegger, M. (1976 [1954]) 'The Question Concerning Technology,' in D. Krell (ed.), *Martin Heidegger: Basic Writings* (New York: Harper & Row), pp. 287–317.

Heidegger, M. (1981 [1966]) '"Only a God Can Save Us": The Spiegel Interview (1966),' in T. Sheehan (ed.), *Heidegger: The Man and the Thinker* (Chicago: Precedent Publishing), pp. 45–67.

Latour, B. (1987) *Science in Action: How to Follow Scientists and Engineers through Society* (Cambridge, MA: Harvard University Press).

Latour, B. (1993) *We Have Never Been Modern* (Cambridge, MA: Harvard University Press).

Latour, B. (2004) *Politics of Nature: How to Bring the Sciences into Democracy* (Cambridge, MA: Harvard University Press).

Olds, J. (1958) 'Self-Stimulation of the Brain: Its Use to Study Local Effects of Hunger, Sex, and Drugs,' *Science, 127* (14 Feb 1958), 315–24.

Pickering, A. (1981) 'The Hunting of the Quark,' *Isis, 72*, 216–36.

Pickering, A. (1984) *Constructing Quarks: A Sociological History of Particle Physics* (Chicago: University of Chicago Press).

Pickering, A. (1989) 'Living in the Material World: On Realism and Experimental Practice,' in D. Gooding, T. J. Pinch and S. Schaffer (eds), *The Uses of Experiment: Studies of Experimentation in the Natural Sciences* (Cambridge University Press), pp. 275–97.

Pickering, A. (1995a) *The Mangle of Practice: Time, Agency, and Science* (Chicago: University of Chicago Press).

Pickering, A. (1995b) 'Cyborg History and the World War II Regime,' *Perspectives on Science, 3*, 1–48.

Pickering, A. (1997) 'History of Economics and the History of Agency,' in J. Henderson (ed.), *The State of the History of Economics: Proceedings of the History of Economics Society* (London: Routledge), pp. 6–18.

Pickering, A. (2002) 'Cybernetics and the Mangle: Ashby, Beer and Pask,' *Social Studies of Science, 32*, 413–37.

Pickering, A. (2004) 'The Science of the Unknowable: Stafford Beer's Cybernetic Informatics,' in Raul Espejo (ed.), *Tribute to Stafford Beer,* special issue of Kybernetes, *33* (2004), 499–521.

Pickering, A. (2005) 'Decentring Sociology: Synthetic Dyes and Social Theory,' and 'From Dyes to Iraq: A Reply to Jonathan Harwood.' In Ursula Klein (ed.), *Technoscientific Productivity, special issue of Perspectives on Science, 13*, 352–405, 416–25.

Pickering, A. (forthcoming a), *Sketches of Another Future: Cybernetics in Britain, 1940-2000.*

Pickering, A. (forthcoming b) 'New Ontologies,' to appear in A. Pickering and K. Guzik (eds), *The Mangle in Practice: Science, Society and Becoming* (Duke University Press).

Pickering, A. (forthcoming c) 'Beyond Design: Cybernetics, Biological Computers and Hylozoism,' to appear in *Synthese.*

18

Daniel Sarewitz

Director Consortium for Science, Policy & Outcomes

Arizona State University
USA

---

1. Why were you initially drawn to philosophical issues concerning technology?

Actually I hadn't thought about this before, but I suppose I am, in fact, "drawn to philosophical issues concerning technology." "When" is easier than "why." In sixth grade—this would have been 1966 or '67—I was the only kid in my class willing to take the "against" position in the class debate about the Apollo mission. I made, as I recall, the opportunity cost argument—why spend all those billions sending men to the moon when there are people without enough to eat in America's inner cities and rural poverty pockets? So here, perhaps, is also a clue about "why." Injustice on all scales has always given me a pit in my stomach, and apparently at the age of eleven I located in technology a key site for the production of injustice.

Several years later, on the other hand, Kubrick's visionary 2001: *A Space Odyssey* opened up for me the grandeur, magnetism, and mystery of space exploration. Perhaps going to the moon and curing poverty were both worth doing. It wasn't until embarrassingly recently, however, that I understood the most important reason why we managed to go to the moon in the 1960s while simultaneously failing to cure poverty—despite the resources devoted to the creation of "The Great Society" in America during that time: The former was easy, and the latter was very hard.

2. What does your work reveal about technology that other academics, citizens, or engineers typically fail to appreciate?

First: The dominant role of technology in the lives of people in affluent societies is to prevent the collapse of the hedonic treadmill that powers market economies.

It seems to be true that continual economic growth is necessary (yes, yes, but obviously insufficient) for continued civil stability in modern market democracies. It is equally the case that continual technological advance is necessary for economic growth. Furthermore, in my view the most profound empirical finding (and one of the most robust, as well) to emerge from the social sciences in recent decades is that subjective quality of life—people's assessments of their own level of well-being, of their happiness—is decoupled from economic growth. Almost all nations for which there are data show incredibly stable levels of subjective well being over the past forty years or so, a period of huge wealth creation: increased societal affluence does not translate into increased feelings of well-being. [1] From one perspective, this is a consoling result: more money really can't buy more happiness! (did we need decades of survey work to prove this?). But from another perspective the detachment of happiness from wealth creation illuminates a troubling incoherence in the relationship between humans and their technology.

Economists of technology like to observe that technological innovation is the dynamo of economic growth, but what they really mean is that the *consumption* of the products of innovation accounts for this growth. Yet when people consume technologies, they are not doing so to make their little contribution to wealth creation, they are doing so to satisfy their own wants (whatever the source of those wants may be).

Indeed, we talk of technology, at its best, as the source of solutions to problems, and we talk of the solution to problems as a key to improving our lives. The converse is also true: because our consumption of technologies leads to the continual remaking of

[1] The reasons are dauntingly obvious: happiness flows from the quality of one's social relations, from deep satisfactions of meaningful work and engaging play. There is no obvious link between more affluence and more of any of these sources of well being. Subjective well-being also flows from a sense of doing well relative to others; more affluence may satisfy this sense for individuals, but not across society as a whole.

the appearance, mechanics, and dynamics of our daily existence—from the way we enjoy music to the way we eat and work and have sex—to imagine ourselves in a previous world is to imagine ourselves in an uncomfortably different world, a world in which most people, presumably, would not wish to live. Give up my i-pod or my portable defibrillator? Forget about it. Yet the data tell us that we don't actually like living in this current version of the world any more than we liked living in previous ones. So our behavior as avid and habitual consumers of wealth-creating and world-transforming technology only makes sense if viewed as society-wide neurosis: apparently we continue to believe that future consumption actually will lead to a better quality of life (or stave off a decline), even though we have never experienced it that way before. Our happiness, satisfaction, and wellness, that is, are coupled to our technological condition only via our vain ambitions for a more satisfying future that never arrives. But if we gave up on technological consumption as the road to a better life then the centrifugal force that holds modern market economies together would dissipate. We are gerbils on an exercise wheel, and our continual running in place is precisely what sustains modern gerbildom.

Second: Technology is a powerful political organizer.

I take it that the essence of a technology is its internalization of one or more cause-and-effect relations and its ability to reproduce those relations with high reliability, often in a wide variety of contexts. A light switch, a thermonuclear device, scuba gear, a vaccine, or the personal computer and software with which I now write: they are remarkably dependable, the embodiments of reliable action. They are also, apparently, considerable improvements over prior technique, e.g., respectively: candles, conventional bombs, holding your breath, quarantine, typewriters.

A lot of writing and thinking about technology in the post-industrial world has focused on its contested nature, canonically displayed in the debates over genetically engineered foods (GMOs) and nuclear power, and, today, human cognitive enhancement. We haven't thought enough about another aspect of technology: its ability to bring diverse political and institutional players into a well-coordinated solar system of collaborative action. If you think about a relatively uncontroversial technology, childhood vaccines for example, the coordination of social activity that has taken place to ensure their effectiveness is quite extraordinary. The vaccine industry, medical practitioners, health insurers, government regulators, school systems, local governments, and individual families have all recognized a common interest that allows them to work together in reasonable harmony to ensure that most children in the U.S. get vaccinated. The gravitational center of this near-miraculous degree of cooperation is the core technology itself—vaccines that yield reliable, and desirable, outcomes, and thus motivate the accretion of the necessary network of actors and institutions, which itself represent an enormous diversity of interests, values, and ways of knowing.

I'm increasingly drawn to the idea that successful technologies can be understood as highly effective agents of political organization. Because technologies are, in effect, reliable cause-and-effect machines, those whose interests and goals are advanced by what a technology does would be foolish not to organize around the use of that technology. Of course those whose interests and goals might be threatened by a given technology may organize against it, and ultimately be disenfranchised by it, but such opposition labors at a huge and inherent disadvantage because the interests of their side are not bolstered by the effectiveness and high reliability of the technology itself. The common image of Luddites is not of fighters for social justice, which they were, but of quixotic, at best, resisters of the inexorable tide of innovation. Opposition to the planting and consumption of GMOs, while currently holding a beachhead in western Europe and Africa, is ultimately doomed because of the highly reliable behavior and delivery of benefits that GMOs offer to their supporters. This high reliability in both performance and delivery of benefits is in strong contrast to the behavior of the variety of actors that needs to be held together by a more abstract commitment to opposing GMOs. Technology stacks the political deck in favor of its own side.

I need to be clear about two aspects of this argument that are likely to be misinterpreted. First, I am not saying that there may not be very good reasons for opposing various technologies because of the potential for social or environmental impacts that adversely affect legitimate interests and important values. I'm only saying that if such opposition is framed in terms of using or not using a technology, those who believe that the technology should not be used are shouldering a grave political and practical disadvantage. Political because, as I am arguing, effective technologies are a much more coherent nucleus around which to organize interests than are abstractions. Practical because effective technologies are, in general, much better at achieving the outcomes for which they are designed than are approaches to problem-solving that do not have, at their core, an effective technology but rather depend, say, on humans behaving or organizing in certain ways. Thus, while I think there is a reasonable basis for concern about GMOs from cultural and ecological perspectives, if the goal is to maximize food production and minimize environmental damage from chemical inputs and land use, then GMOs are likely, on the whole, to define a more successful path than approaches that depend for success on complex, integrated changes in the behavior of diverse actors in the global agricultural system.

Nor am I making an argument on behalf of technological determinism. We all know that technologies and the complex social and institutional networks that grow up around them are products of social systems and human choices. Nevertheless, to fail to recognize the action essence of technology—it's reliable embodiment of cause-and-effect relations, of effectiveness in the world—is to fail to understand the power of technology, in turn, to influence those systems and choices. One might object that the notion of "technological effectiveness" is incoherent because such effectiveness does not exist in a vacuum but demands complicit organization of social systems. But powerful positive feedbacks emerge between such systems and the technologies at their core. If one chooses to alienate one's self from the technologies, then one also isolates one's self from the feedbacks and the powerfully effective actions they enable.

So I've made two points here about technology that I think are insufficiently incorporated into thinking about how society works, and in particular into discussions about technology and positive social change. These points may seem to be contradictory, or at least in tension: first, that technological change in market democracies is a Sisyphean project when it comes to improving people's own sense of well-being, and second, that technologies themselves are powerful engines of social action and, therefore, political organization. These points may appear to be contradictory because it seems sensible that improved life satisfaction would naturally be one measure of the worthwhileness of social action. That is, if technologies embody effectiveness at achieving the ends desired by those who deploy them, wouldn't one expect that they would also add to the satisfaction of those people as well, through the process of fulfilling desires? So one way of interpreting things is to say that the contradiction is real, that technology obviously is at best only weakly coupled to the achievement of life satisfaction, and therefore that we need to look to other, non-technological sources of public and private value in the search for a path to greater and more meaningful well being. I take this as a pretty widely accepted perspective among political progressives in an era where the threats, failures, and uncertainties of technology are manifest. I am hugely sympathetic with it. But it raises a very troubling problem: if progressives don't take technology seriously, then those that do will have an immense operational and political advantage in the world.

3. What, if any, practical and/or social-political obligations follow from studying technology from a philosophical perspective?

So here's an example from the real world: an academic biomedical researcher with an economic stake in a particular genetic therapy wants to test it on a patient. A bioethicist at the same university is consulted and approves of the experiment (I presume, perhaps wrongly, that bioethicists are supposed to engage in philosophical reflection?). Patient dies, but the scientist has fulfilled the obligation that follows from studying technology from a philosophical perspective, right?

At the opposite end of the spectrum we have the precautionary principle, two words which I have a really hard time typing out or saying without first appending the disclaimer "so-called," and which is probably the crowning jewel (or maybe the only jewel) of philosophical "obligations" relating to the governance of technology. Of course I understand that in its generic version the PP just means that, given uncertainties about the health and environmental risks of new technologies, erring on the side of prudence is a good thing, as is providing a regulatory and psychological counterbalance to the exuberance of profit-motivated technology producers. Who could argue with this, until you try to really figure out what "erring" and "prudence" mean when you're talking about complex, networked, technological systems in which cause-effect relations are pretty much impossible to parse. For example, how easy is it to know what the truly precautionary position is for using DDT against malaria in Africa?

I've argued, in essence, that rejecting technology is rejecting power. If you're serious about your politics, then this is a bad place to begin. It seems odd to me that progressive politics has been, on the whole, willing to adopt science as a legitimating source of cultural power while maintaining a high degree of skepticism about technology. Science is inherently contestable and open-ended, and thus equally available to all parties. It is an ineffectual tool for political persuasion. It also does not translate easily into action. Technology, on the other hand, can short-circuit political conflict not by persuading diverse interests and values to converge, but by allowing them to co-exist in a shared sense of practical benefits derived from the action embodied in technology. A great example is the story of how the nations of the world came to an agreement in the late 1980s to phase out the production of chlorofluorocarbons—a technologically and economically important refrigerant and solvent that also happened to destroy the stratospheric ozone layer that protects Earth from harmful ultraviolet radiation.

Popular narrative has it that the CFC-ozone problem was solved when the science proved that CFCs had caused the Antarctic ozone hole, and the world, faced with definitive knowledge, took effective action. This is a story of science forcing right behavior, but it is bizarrely incomplete, because it ignores the question of what "effective action" actually means, and how it is to be taken. The many nations that agreed to phase out CFC production did not, in so doing, decide to live without the benefits of keeping refrigerators cold. The missing element here is the technological. Neither the politics nor the science surrounding the ozone debate was settled until Dupont invented a CFC substitute, hydrochlorofluorocarbons, that was less destructive of ozone. The availability of an alternative to CFCs made it possible at once to meet the goals of three distinct constituencies: those whose primary interest was to protect the ozone layer, those whose primary interest was to make money producing chemicals, and those, especially in the developing world, who were unable to give up on the benefits that CFCs alone could provide in an economically viable way. This story is perhaps less satisfying than the tale of science convincing people to make sacrifices for the good of the planet and humanity, to do the right thing regardless of worldly consequences, but it has the virtue of actually explaining how effective action was able to come about. Sometimes a technological fix is a beautiful thing.

A predisposition toward rejecting technological approaches to social problems—PP is rooted in the predisposition—fosters impotence in two ways, first by rejecting the worldly effectiveness of technology, and, second, by doing without the political organizing value created by such effectiveness. This phenomenon is on depressingly stark display with the problem of climate change, where the political agenda has focused on using science to compel political action, with the result that we've got lots of science combined with a political agenda that has had no discernible effect on preventing either climate change or damaging climate impacts. Attributing this failure to lack of "political will" simply restates the problem, since "lack of political will" is just a synonym for "not wanting to do things my way." Climate change will have many solutions, but the transition to a carbon-free energy economy must largely be a technological one, so if progressives are really serious about climate change, then technological systems should be central to their political agenda. Why haven't they been? I find it really difficult not to think that the political framing of the climate change problem is largely impelled by concern about the moral meanings of affluence (excessive consumption and the like; return to question 2, part 1, above) rather than by concern about the consequences of climate change itself. Take this test: if in the next decade or two technologies for capturing and storing carbon emissions create the realistic prospect of essentially decarbonizing the global energy system, would you be satisfied, or would you feel somehow cheated of the opportunity to make people work harder for their redemption? Where does obligation most potently lie? On the side of trying to get people to behave differently, or of trying to advance technologies that reduce the need for behavioral change?

4. If the history of ideas were to be narrated in such a way as to emphasize technological issues, how would that narrative differ from traditional accounts?

I know very little about the history of ideas, and even less about what constitutes a "traditional" account. Given received myths about inexorable and inevitable progress in the world (and their antithetic myths of the coming eco-socio-techno-dystopia), it's not clear to me that the problem is lack of emphasis on technology but lack of clarity about how our philosophies and our normative commitments relate to our technological narratives.

5. With respect to present and future inquiry, how can the most important philosophical problems concerning technology be identified and explored?

In the past year I've heard compelling philosophical treatments of both the moral obligation to rapidly pursue human enhancement technologies and the obligation to abandon such pursuit. I've heard a neuroscientist, fighting back after a series of philosophical and political assaults on enhancement research, ask why science aimed at helping people overcome their cognitive disabilities and limits should be approached from a precautionary perspective but not energy research aimed at helping the world overcome its dependence on hydrocarbons? It's a perfectly reasonable question. I don't know much about philosophy, as is probably obvious, but my observation is that the ethics of justifying biomedical technology and the ratiocinations underlying the precautionary principle, while rooted in opposite assumptions about technology (one permissive, the other conservative), each address a decontextualized and therefore non-existent world where individual agency and artifacts are the proper unit of analysis for addressing the question of what we ought or ought not to do regarding questions of technology. Such approaches, on both the optimistic and pessimistic sides, seem, to me, to fail to engage the fact that innovation is as defining a human attribute as just about anything else you can name, and that the human prospect just cannot be extricated from the technological systems that we humans have created and on which we entirely, now, depend.

If technological innovation is understood as a core activity of the human species, as organic as composing music or falling in love—which it is—rather than an elective hobby that can either be pursued or not, then the core question about technology becomes one of governing, of modulating, the innovation activity itself. We ought, it seems, to be striving for a philosophy of the innovation process, of the role of technology system evolution in human affairs, rather than a philosophy of the creation and meanings of technological artifacts. We know a lot about how innovation occurs, especially its decentralization among institutions and sectors and actors. It's a networked process, it's usually incremental, it's not cleanly separable from other social activities and institutions. Human choice is present at many venues, some of which provide greater leverage than others, but none of which allow significant control, in that there is no site that can visualize a future technological system and make it happen according to plan. People who think a lot about technology and society understand all this quite well, and the key problem of intervention that it entails: the better we can see and understand a technological system, the more difficult it is to have much impact on its evolution, both because of the inertia, or lock-in, that gets built into an operating system (e.g., the problem with automobiles is not just internal combustion engines, it's also roads and filling stations and suburbs, the organization of manufacturing, etc. etc.), and because of the increasing economic and political power of those whose interests are served by the technology. But at earlier stages of evolution, the technologies and interests surrounding them are insufficiently formed to motivate action or to permit accurate prediction of how the systems will evolve.

From a political and policy perspective, the solution to this dilemma is now more-or-less clear. The innovation process needs continually to be subjected to the rigors of pluralistic reflection and discourse. This process needs to take place throughout the complex, disseminated venues where decisions related to innovation are made: laboratories, government research agencies, legislatures, corporate board rooms, retails stores. The contexts from which innovation arise need to be as self-conscious, argumentative, and openly normative as any other important social activity with world-transforming potential.

For me, the most compelling and worrisome technologies on the horizon now are those having to do with the enhancement of human cognitive function. Advances in fields ranging from neuroscience to nanotechnology to genomics are creating the potential to increase memory, expand sensory and information processing capabilities, and even enable direct, remote communication between brains and machines, and between one person's brain and another's. Technical experts talk confidently about the potential to enhance intelligence, they also raise the possibility for improvement of intangible qualities like judgment and wisdom—and even subjective well-being, or happiness.

There are plenty of good reasons to squirm at these prospects: the potential for using such technologies to diminish the cognition of our enemies; the injustice and inequity that would flow from inevitable problems of unequal access; the fact that some of the attributes said to be enhanceable, like intelligence, are in part socially constructed and thus internalize existing social power structures; the profound challenge that such technologies could create for the functioning of our fundamental democratic institutions, and even for economic markets, which have arisen, organically as it were, from a species with a particular range and distribution of cognitive assets.

But an approach to cognitive enhancement that takes rejection as a starting point is an untenable foundation for action in the real world, and is instead a recipe for marginalization. Because the boundary between therapy for those who suffer from cognitive impairment, and enhancement for the "normal," is fuzzy, the moral case for complete rejection is a difficult one to make. The practical case is impossible to make: in an increasingly globalized technological innovation system, the possibility for suppression of effective technologies that can advance powerful interests is nil, as we see in the continued proliferation of nuclear weapons and GMOs. And if some of these enhancement technologies prove effective, they will instantly attract constituencies who, in benefiting directly from the enhancements, will thus be better able to act more effectively in the world. Transhumanism, the small but loud cultural movement rooted in the idea that part of what makes us human is our continual pursuit of "better" human performance through technology—I enhance, therefore I am—is a force to be reckoned with not because of its tortured philosophizing but because it aligns its own interests with technological innovation.

To bring things to an end, I must confess to a smidgeon of optimism here. It actually does seem to me that public discussions surrounding emerging areas of world-changing innovation, such as nanotechnology, genomics, and cognitive enhancement, are more intense, and are penetrating more deeply and more pervasively into the innovation process, at an earlier stage, than has ever been the case. The site of philosophical engagement here needs to move away from the specific technologies (most of which don't yet exist and at least some of which may never prove feasible) to the dispersed components of the innovation processes from which complex technological systems will emerge. Innovation is a political activity, which means that there is no optimal outcome. Agency is not really located in individuals, so standard moral yardsticks for human behavior are not particularly relevant. If the principles upon which a good society must be built—justice, tolerance, equity, sustainability, etc.—are applied only as the measures of technologies themselves, then they will be applied impotently, and too late. Rather, they must inform the initial conditions of human choice from which innovation, in all its diversity and unpredictability, can proceed.

Additional Readings

Sarewitz, D., G. Foladori, N. Invernizzi, and M. Garfinkel 2004. Science Policy in its Social Context, *Philosophy Today, v. 48(5):* 67–83.

Sarewitz, D. 2003. Science and Happiness, in: A. Lightman, D. Sarewitz and C. Desser (eds.). *Living with the Genie: Essays on Technology and the Quest for Human Mastery*, Washington, DC: Island Press, pp. 181–200..

Sarewitz, D. 1996. *Frontiers of Illusion*: *Science, Technology and Politics of Progress.* Philadelphia: Temple University Press.

19

Evan Selinger

Department of Philosophy

Rochester Institute of Technology
USA

---

1. Why were you initially drawn to philosophical issues concerning technology?

Although it was risky, I came to college wanting to major in philosophy. Having been on my own, financially and legally, since the age of 16, it would have been practical to study accounting or law. But the chance exposure to Friedrich Nietzsche while I was in High School was too powerful to be contained. Even before taking Introduction to Philosophy, I had a sense that philosophy grappled with issues that everyday consciousness neither acknowledged nor had the conceptual tools to reckon with. Given the disparities between how my upper-middle class Long Island peers and I looked at the world, I entered college hoping that philosophy would help me better understand alienation, including my own personal experiences of it.

Given this background, I initially gravitated to the "Continental" traditions of philosophy that condemned technology's oppressive consequences. The principals I studied wrote with dystopian flair, claiming that modern technology serves oppressive ends, and that such oppression goes largely unnoticed as a consequence of technology being treated as value-free, and a result of it routinely being associated, benignly, with human nature, cultural innovation and reproduction, and medical benefits.

My exposure to Martin Heidegger's writings convinced me that "Western technological thinking" culminated in an epidemic loss of creativity and basic humanity. Heidegger's central message was that the human capacity to build new "worlds"—that is, to create innovative understandings of reality—has been so thoroughly stymied through the seductions of efficiency, that meaningful change could only come from the grace bestowed upon us by a *Gestalt* switch, i.e., a new understanding of Being that he enigmatically identifies with a redemptive "god."

Similarly, I was persuaded by the disenchanted view of techno-capitalism that the early Frankfurt School Critical Theorists (abbrev. FSCT), including Theodor Adorno, Max Horkheimer, Herbert Marcuse, and Hannah Arendt, developed. Prior to reading them, I thought that Nazi fascism and free-market enterprise had little in common. But these philosophers turned bureaucracy and the bureaucratic mentality into proper topics of philosophical inquiry; in so doing, they provided compelling reasons to reconsider matters.

At face value, bureaucracy appears to be little more than an occasionally frustrating structural feature of the modern division of labor. Without it, goods could not be efficiently produced, distributed, or exchanged. But according to FSCT, bureaucracy and the bureaucratic mentality are also forces that participate in the eradication of personal and collective responsibility. Primarily, this is because the precisely defined procedures that typify bureaucratic institutions substitute hierarchy and general rules for personal judgment; accountability, therefore, is deferred through chains-of-command, paperwork, and precedents that can be outdated and too expensive to change. In the final analysis, according to FSCT, a range of bureaucrats, including the "banally evil" Nazi soldier as well as the contemporary corporate drone, lack the capacity to exercise genuinely responsible action: Each one of them follows orders, and avoids the burden of accountability by taking refuge in the fact that their behavior is rigidly determined.

While technology may not be the only cause of this alleged moral vacuum, FSCT argue that modern bureaucratic systems would not have their all-encompassing character were it not for the proliferation of specialized labor and regulatory techniques that inhibit critical reflection and obstruct emancipatory social relations. And, they further stigmatize technology by emphasizing the fact that the false needs of "commodity fetishism" are produced, distributed, and advertised through modern technological means of production.

As my answers to the remaining questions in this interview attest, I still benefit from considering the views outlined above. However, I no longer see them as definitive for philosophizing about technology. My change of position came about as a consequence of being exposed to Don Ihde's *Phenomenology of Technological Ex-*

perience, a philosophical approach to "human-technology-world" relations that emphasizes *multistability* and *ambivalence* by detailing how technological practices simultaneously amplify and reduce perception, as well as simultaneously constrain and augment capacities for thinking and acting. To be sure, I still grapple with the fact that some of the suffering referred to by the term "alienation" abounds. But now I'm convinced that, at least within the context of many contemporary technological practices, *alienation tends to occur, but only partially, incompletely, and often in conjunction with a range of life-affirming possibilities.*

Given this change of perspective, I would argue that a primary task for philosophers of technology is to *render the complexity of technological experience explicit.* While it may seem politically radical to level sweeping critiques of technology, such polemics put theorists at risk for conflating description and projection; far from being accurate depictions of real states-of-affairs and events, ascriptions of "one-dimensional" behavior, epochal understandings, and zero-sum situations tend to be overly-reductive analytic constructs, indicative of normative passions trumping careful description. Thus, as counter-intuitive as it may initially seem, our capacity to think about and build better worlds to live in may very well depend upon our capacity to challenge common perceptions of technology by adding, rather than reducing, the actors, actions, and relations that deserve consideration![1] Of course, the issue at hand isn't simply about expanding domains of analyses; as my answers to the third, fourth, and fifth interview questions suggest, the problem of selecting the right language with which to conceptualize technological experience remains daunting as well.

2. What does your work reveal about technology that other academics, citizens, or engineers typically fail to appreciate?

I'll answer this question by elaborating on a theme broached in the last question—the difficulty of describing technological experience without importing biased projections that distort how the subjects

[1] Despite Bruno Latour's polemic against phenomenology, I believe that this expansive conception of the philosophy of technology establishes common ground between his actor-network approach to science and technology studies and Ihde's. Indeed, sharing some of Latour's reservations about traditional approaches to phenomenology, Ihde now characterizes his own work as "postphenomenology."

of analysis are represented. Here, I'll comment upon aspects of my work that challenge descriptions of experience proffered by one of the most famous living phenomenologists, Hubert Dreyfus.

Dreyfus gave applied phenomenology new life by offering up provocative criticisms of artificial intelligence and expert computational systems. It is fair to credit him for facilitating fruitful interdisciplinary exchanges between phenomenology, computer science, and information technology; and I have even appealed to Dreyfus's phenomenology of embodiment when debating Harry Collins about his sociology of expertise. As I see it, Collins could improve the quality of his research if he were more sympathetic to Dreyfus's basic understanding of the relation between human embodiment and the human capacity for skill acquisition. Indeed, drawing from Dreyfus's philosophy, I have argued that Collins not only misdescribes expert skill, but also risks characterizing handicapped human beings in terms that are more appropriate to impute to computers than humans.

Despite these merits, I find that Dreyfus presents untenable descriptions of experience when he attempts to differentiate how digital computers process information (or, perhaps, uninterpreted symbols) from the ways in which humans perceive and act. To some extent, Dreyfus's descriptions are skewed because he imports Heideggerian biases concerning "authenticity" and "inauthenticity" to his portrayal of experts as an "endangered" species. For Dreyfus, the main problem that experts face is that they are embedded in a culture permeated by inauthentic norms. In other words, expert judgment typically is not considered trustworthy when it is conveyed in the "authentic" language of demonstratives, i.e., propositions that convey aspects of a particular situation which can only be fully appreciated by other experts from the same domain. Instead, to gain public credibility, experts often need to justify their conclusions and recommendations by appealing to "objective" standards, i.e., publicly available rules and commonly accepted procedures for applying these rules to particular cases. As a consequence of the demand for "inauthentic" discourse, Dreyfus laments that human experts are being pressured to refrain from engaging in the type of action that typifies humanity at its apex of skill. What such pressure discloses, according to Dreyfus, are the detrimental social consequences that have arisen, in part, due to widespread endorsement of the computational conception of consciousness: A leveling effect occurs in which the threshold separating both inauthentic human non-experts and computers from authentic human experts is obscured; experts and non-experts alike are subject to process of social discipline in which algorithmic protocols for decision-making are valorized and the wisdom of contextually specific judgment is demeaned as mere subjectivism.

In light of these background presuppositions, Dreyfus is driven to present an account of intuition that demarcates skilled human perception from rule-governed computational analysis of data. Unfortunately, such a drive inclines Dreyfus to generate a philosophy of human expertise that distorts fundamental relations between embodiment and embeddedness. In an effort to restore our appreciation of authenticity, Dreyfus goes too far; he obscures the ways in which social biases can influence a human expert's perceptual *Gestalts*, and he thereby romanticizes how human experts develop and use intuition.

In Dreyfus's analysis, experts, including natural scientists, are consistently described as trustworthy people who, *qua* expert perceivers, invariably engage in skillful acts of perception that are impervious to contamination by desires or prejudices originating in experiences that fall outside the scope of the proper domain of their expertise. This is how Robert Crease and I put it in a co-authored article: "From Dreyfus's perspective, one develops the affective comportment and intuitive capacity of an expert solely by immersion into a practice; the skill-acquiring body is assumed to be able, in principle at least, to become the locus of intuition without influence by forces external to the practice in which one is apprenticed."

Such a view is out of touch with the literatures that detail the complex domain-crossing and domain-redefining activities that typify many controversies surrounding expert judgment. And, Dreyfus's position can be interpreted as a tacit endorsement of all expert authority originating from experts who possess a successful track-record of past performance. Contrary to Dreyfus, the experiential gaps that separate past exercises of expert judgment from current ones call into question the soundness of confidently associating track record with trustworthiness. In these gaps, the frontiers of knowledge can shift, the motivations underlying expert judgment can change, and the context for rendering expert appraisal can vary.

The problem of describing human perception in idealized terms extends to Dreyfus's depiction of experts who are neither scientists nor engineers. Despite the prevalence of contrary testimony,

Dreyfus disembodies his account of the experience of playing chess in order to describe a special class of experts, chess Grandmasters, as lightening-quick pattern perceivers who, *qua* Grandmasters, are immune to the distractions of social forces and the seductions of erotic yearnings. Such an idealized account of being in the "zone" is especially problematic for Dreyfus; he treats Grandmasters as a paradigm case that exemplifies fundamental differences between expert human perception and computational approaches to analyzing data. The irony, here, is that even as a pre-eminent theorist of embodiment, Dreyfus reinforces the longstanding but incorrect prejudice that chess is merely a mental activity—albeit a mental activity that symbolizes the pinnacle of human cognition.

Again, I'm placing emphasis upon Dreyfus's philosophy of technology because it illustrates a general problem, one whose scope extends far beyond the particulars of Dreyfus's own philosophy. The main problem at issue concerns how humanist bias—that is, the bias to discern a standard that fundamentally demarcates humans from computers—can prevent philosophers from grasping the complexity of human-technology relations. Because Dreyfus studies the *limits of computational development without also attending to the limits of humanist thinking, he allows polarizing contrast classes (humans vs. computers and authenticity vs. inauthenticity) to bias his putatively value-free descriptions of lived human experience.* Contrary to Dreyfus, I would argue that a critical perspective on technology necessitates a *symmetrical course of study*: it can be difficult to discern the limitations of technology without idealizing human attributes, skills, and standards; and it can be difficult to discern the significance of outputs generated by machines, as well as the benefits that arise when humans use technology, without understanding core human abilities and values.

3. What, if any, practical and/or social-political obligations follow from studying technology from a philosophical perspective?

This question gets at a longstanding controversy—whether philosophical insight can justifiably shatter "false consciousness," or whether the imputation of false consciousness is an intellectually self-serving projection: in effect, a question of whether philosophers are patronizing the so-called masses in order to feel superior to them. On this issue, Bruno Latour presents insightful analysis of the limits of "ideology critique." Latour's intervention should give philosophers pause for thought before they proclaim that the "*hoi polloi*" are mystified and in need of having their unbenighted status uplifted through iconoclastic philosophical wisdom.

Putting Latour's reservations aside for the moment, it should be acknowledged that philosophers of technology have indeed addressed issues that dramatically impact how people think and act. Consider, for example, the new book by Peter Singer and Jim Mason, *The Way We Eat.* Singer and Mason contend that eating is a moral activity because the food humans consume is prepared through practices that impact how animals, the environment, and other human beings are treated. In this context, the authors provide a rich examination of farming and fishing in order to show that a gap exists between what typical consumers imagine occurs with these activities and what actually occurs as a result of specific technological procedures. This emphasis upon the ignorance that separates widespread understanding of how production and consumption are linked has a distinguished philosophical lineage. For example, it has a forceful presence in the writings of Karl Marx's critique of the reliance of industrial capitalism upon factory labor, and it animates Albert Borgmann's meditations on the differences between how objects are perceived and used as "devices" and "focal practices."

Returning to Singer and Mason—

When they vividly detail the technological and economic forces that give rise to factory farming and the procedures that become normalized therein, they clarify how the distance between the means and ends of production impact consumer attitudes in morally questionable ways. Based on anecdotal evidence, this approach appears to have greatly influenced how many people view vegetarianism and "conscientious omnivorism".

However—

Because factory farming is an extreme practice—one so severe that it exemplifies Heidegger's concerns about the essence of modern technology reducing everything to "standing reserve"—it may not be surprising that philosophical analysis can impose new obligations upon those who are just coming to understand factory farming's restrictive forms of operation. From my perspective, provocative normative questions also need to be raised about more fundamentally *ambiguous technological situations*—situations where personal and collective transformation occur, but for which preference utilitarianism and other traditional moral perspectives are not the most comprehensive frameworks of analysis to apply. *In these contexts, the main issue at stake isn't the gap between production and consumption; rather, the difficulty concerns how best to interpret and assess technological experience.*

In my recent work on globalization and development, I have tried to discern whether a more nuanced understanding of "empowerment" can improve contemporary assessments of technologically-oriented micro-credit programs, such as the Grameen Bank's Village Phone program (VP) in Bangladesh. This topic returns us to the main issue discussed in response to the first interview question—that is, the conceptual value of complicating dominant conceptions of human-technology relations.[2]

In summary, VP emerged from initiatives taken in 1997 by Muhammad Yunus, an economist who subsequently won the 2006 Nobel Peace Prize. At that time, the Grameen Bank began a collaborative venture between two companies, a private for-profit company, GrameenPhone Ltd., and a not-for-profit one, Grameen Telecom, to apply micro-credit principles to assist impoverished women so they could acquire mobile phones, a commodity that could be rented out to fellow villagers on a call-by-call basis. Given the limited number of existing landline phones, coupled with the prohibitive expense that prevents impoverished Bangladeshis from acquiring their own mobile phones, this plan—which also includes an explicit liberal social agenda ("Sixteen Decisions")—appears to promote human rights through economic reform and technology transfer. The GrameenPhone website earnestly characterizes the telephone as a "weapon against poverty," and the women who participate in the program as "phone ladies."

One of the central debates concerning this project, however, concerns whether or not Bangladeshi women actually become empowered by participating in it.

Typically, five reasons are given in favor of viewing VP as an empowering program. First, in targeting women directly, programs like VP are said to empower the "poorest of the poor." Second, VP is credited for giving women opportunities for employment that traditional Muslim customs of *purdah* inhibit. Third, as a consequence of these economic opportunities, women are said to

[2]In making this claim, I am not suggesting that the relation between production and consumption is irrelevant to understanding and assessing projects like the Village Phone program. To the contrary, that relation is significant and it needs to be better understood. My point is simply that the relation between production and conception is not the main issue at stake, descriptively or normatively.

acquire more authority over household decisions; they gain new respect in their communities and with their spouses. Fourth, with increased income, women are said to be capable of taking a more active role in their children's futures, e.g., instilling a more positive image about women to their daughters, and having resources to provide their children, boys and girls, with better educational opportunities. Fifth, due to the Grameen Bank's "Sixteen Resolutions," women are said to be eschewing the repressive custom of dowry and learning skills that instill self-discipline and appreciation for wellness (nutrition and sanitation is emphasized) as well as solidarity (pledges to look after one another are made). Sixth, by promoting "entrepreneurialism," micro-credit is said to do something that charity cannot—instill pride and confidence, characteristics that can form the foundation for an improved political participation.

Contrarily, some critics claim that VP is fundamentally a disempowering program that "co-opts" and "appropriates" women's struggles by reducing women to "welfare objects" of reform that are given little "ownership" over the programs they participate in. Allegedly, husbands, development workers, and paid professionals remain the main authorities over phone ladies' lives. In this context, it is said that women are patronized, i.e., viewed as "incapable" of "identifying their own needs and priorities," and unable to exercise their own "rationality" to develop positive "strategies" and "visions" for uplifting their personal and collective situations.

Additionally, critics charge that programs like VP engage in insidious acts of social discipline by instilling the capitalist ethos of "individualism" and consumerism into women's psyches and bodies. Individualism is said to be promoted because phone ladies accrue new familial and social opportunities through competitive commercial practices. Not only are such practices said to inhibit opportunities for collective consciousness to form, but they appear to even co-opt the ideal, e.g., the "solidarity" circles that the Grameen Bank relies upon are alleged to be calculated mechanisms designed to minimize the transaction cost for ensuring that women repay their loans.

As to commercialism, women are rewarded for believing that money is a proper source of respect. By slightly improving the material conditions of individual women—with what is, only comparatively speaking, better material conditions—micro-credit programs are said to provide an incentive for women: (1) to avoid re-examining the systemic sources of their collective oppression, and (2) to elevate instrumental values of efficiency and purchasing power to personal virtues.

These critics are ultimately insisting that empowerment, properly defined, is the capacity of agents to actively discover their capacity to make principled choices and to create environments in which they can express themselves politically through symbolic action.

What makes this debate so difficult is that discussing it through the prism of traditional moral prohibitions and licenses will not clarify matters. VP is not so insidious that it should be rendered illegal. And yet, it may be producing certain consequences and harms that call for more criticism than would be warranted under a libertarian framework.

My main contention is that both critics and advocates of VP typically work within the parameters of *overly modernist conceptions of empowerment and disempowerment.* What this binary conceals is that the experience at issue entails *simultaneous relations of both dependence and independence.* Like Hannah Arendt in *The Human Condition*, the critics place too high a premium on symbolic efforts that fall outside the bounds of technological and economic reform. In this respect, their humanism is patronizing; the disqualification of technology and economics from the domain of politics results in the reduction of experiential claims about empowerment to false consciousness. Advocates, however, do not fully grasp how techno-economic programs create unintended consequences and ambiguous results that their theoretical analyses fail to register. In this respect, technocracy generates overly idealized visions of programmatic success and undue confidence in the reasons for program replication.

Given the limits of this binary, I'm working to revise the discourse on empowerment in a way that emphasizes the *emergence of simultaneous relations of independence and dependence.* Understood in this way, it becomes clear that in order for phone ladies to gain some economic and social independence, double-dependence is required. The phone ladies are dependent on following a techno-economic script; without that script, they fail to gain respect. And, the villagers are rendered dependent on limited access to a technological commodity. Thus, if a new economic program comes around that makes mobile telecommunications more affordable for personal consumptions, the phone ladies risk losing their social stature.

In short, I'm contending that the phone lady experience of empowerment is a "hybrid" of independence and dependence. Programs like VP can create independence only by capitalizing on, and possibly perpetuating, a variety of dependency relations. Until this thought is fully understood, I believe that inadequate standards for assessing its success will continue to be applied.

I recognize that in answering this question, I may not have articulated strict obligations that theorists or practitioners should follow. Instead, I hope to have provided a new perspective on a technological issue that has social and political consequences. Ideally, at some point in the near future, I'll be in greater dialogue with people working in the development sector.

4. If the history of ideas were to be narrated in such a way as to emphasize technological issues, how would that narrative differ from traditional accounts?

I'll answer this question by continuing to discuss the difficulty of selecting appropriate concepts for analyzing technologically mediated experiences of globalization. Whereas my answer to the previous question focused on the problem of extending modern concepts to what Bruno Latour calls "amodern" situations, here I will turn to another hermeneutic issue—using non-traditional concepts to unduly demean political struggles over technology.

In Thomas Friedman's popular book, *The World Is Flat*, he refers to the new mobile and digital technologies as the "steroids" of globalization: They are, he says, beginning to allow individuals to shape, manipulate, and transmit information at very high speeds and with total ease "from anywhere, with anyone through any device" in the world. By "steroids," Friedman means that such technologies "amplify" and "turbocharge" the central forces of contemporary globalization: "outsourcing, offshoring, uploading, supply-chaining, insourcing, and in-forming." Although these "steroids" currently "are caught in a maze of wireless technology offerings and standards that are still not totally interoperable," Friedman envisions the completion of the "mobile me" revolution as occurring sometime in the near future—a time "when you can move seamlessly around the town, the country, or the world with whatever device you want." "When this is fully diffused," Friedman declares, "the 'mobile me' will have its full flattening effect, by freeing people to truly be able to work and communicate from anywhere to anywhere with anything."

Although Friedman discusses a variety of technological, economic, and political factors that have led up to the "mobile me" revolution, his account can be understood as insufficiently ontological. By restricting his narrative to chronological, empirical history.

Friedman never questions whether the human capacity for becoming integrated into a dynamic world of mobile telecommunications technologies is a capacity that lies at the very heart of human nature. In contrast, Andy Clark provides just such an account in *Natural Born Cyborgs: Minds, Technologies, and the Future of Human Intelligence.* In that work, Clark revisits the theme of *homo faber* and in doing so provides new conceptual grounds for reconsidering many aspects of empirical as well as intellectual history.

As an update to his earlier "extended mind" thesis, Clark defines human beings as "cyborgs." According to Clark, humans have always been "reasoning systems whose minds and selves are spread across biological brain and nonbiological circuitry." He insists that collaborative and literal mergers with external, nonbiological tools have always been a way for humans to be "cognitive opportunists." Such mergers allegedly allow us to extend cognitive and problem-solving capacities beyond the biological "skinbag," a threshold that includes the inherited architecture of the human mind. Framed in this way, Clark depicts bio-technological mergers with contemporary technologies, including the ones identified by Friedman as steroids, as merely following from and extending "natural" stages of human development.

Clark's account of cyborgs is useful for rethinking narratives about intellectual history because it presents compelling reasons to re-evaluate dominant portrayals of the invention of ideas. For example, Clark contends that in order to fully grasp the historical significance of meta-cognitive discourse—a communally available form of thought that can address metaphysical, epistemological, and normative issues—it is crucial to consider how the human experience of collective reasoning required the invention of inscription technologies. What Clark demonstrates, therefore, is that a comprehensive account of the history of ideas is one that asks the following sorts of questions: What material practices allow ideas, including philosophical ones, to be generated, disseminated, and reconfigured? How do changes to these material practices influence the type of ideas that are generated and the style in which they are presented, or influence the processes of dissemination and attribution, and the norms for intellectual refutation and extension that emerge?

Despite these merits, Timothy Engström and I have concluded that Clark's cyborg philosophy exemplifies *a general problem that all philosophers of technology need to be wary of.* Clark tends to treat some of the dominant buzz-words found in today's digital marketplace as appropriate philosophical terminology for his "naturalist" ontology, one that is said to underlie *historical and contemporary* human experiences of technology that traditional philosophical theories have not adequately grasped. In projecting backwards, however, from the cyborgian to the natural Clark ignores some of the corrosive effects of marketing feedback loops:

- The more inclined we are to see our personal and collective desires through the images and vocabularies that marketers promote, such as "upgrades," "enhancements," and "opportunism," the less inclined we may be to question the adequacy of those very images and vocabularies as signifiers of human potential, human solidarity, and human flourishing, let alone a representation of "human nature."

- The more inclined we are to equate technological advances in information management with improved social relations, the less inclined we may be to confront what Daniel Sarewitz (in his interview for this volume) calls the "neurosis" of the hedonic treadmill.

- The more inclined we are to equate technological advances in information management with improved social relations, the less inclined we may be to question the social and political implications that follow when information technologies are unequally distributed, or to consider issues of nation, class, gender, and race that might rightly arise as a result of their distribution.

- The more inclined we are to treat commercially incentivized, individually expressed "choices" of technological procurement and application as synonymous with democratic participation, the less inclined we may be to question whether uniform choices about technological consumption are truly sanctioned by or beneficial to a normatively legitimate conception of the public and its collective and deliberative concerns.

- The more inclined we are to treat technological consumption as a fundamental extension of human nature and an enhancement of human cognitive capacity, the less inclined we may be to view technological consumption as itself a political issue.

- The more inclined we are to see human progress as being fundamentally guided by our capacity to promote and symbiotically appropriate technological innovation, the less inclined we may be to view human progress in terms of resource conservation and distributive justice.

It can come as no surprise, therefore, that a market researcher such as Markus Giesler would actively revel in the elision of commercial and ontological discourses. He is essentially suggesting that leading technology companies, including Apple, continue to be financially successful precisely because they already recognize, even if only implicitly, the allure of cyborg discourse.

5. With respect to present and future inquiry, how can the most important philosophical problems concerning technology be identified and explored?

Because the pace of innovative technological and scientific imaging practices are outstripping our capacity to understand them, I would like to see improved cooperation between philosophers of science and technology, philosophers of art and aesthetics, and social-political philosophers while they cooperatively address the theme of digital information and the "visual turn."

In particular, critical attention needs to be given to the topic of simulation. For by turning to simulation, it becomes possible to make headway in discerning the extent to which digital imaging has ushered in a new range of visual practices whose implications simultaneously remain tied to longstanding material and conceptual histories and also exceed the consequences associated with prior technological and visual revolutions.

Simply put, not only do we lack an adequate understanding of what type of science simulation is (or, perhaps, what types of sciences simulations are), but we also have yet to comprehensively address the social-political consequences of living during a time in which computational simulations are increasingly becoming commonplace in technical and scientific domains of research, educational theory and practice, and public policy initiatives involving the manipulation of natural environments, material goods, human labor, and the distribution of social populations. Here, too, important questions need to be raised about diverse forms of contemporary aesthetic and representational practices, ranging from media and experiences that revolve around moving images, such as cinema and video games (particularly games that simulate real military conflicts), to the less pictorial forms of expression explored through the literary imagination. On this last issue, it may be more instructive to examine the emergence of new hybrid and hyper-literary forms of writing than to decry the putative demise of the book in the digital age.

In conjunction with these trajectories of inquiry, renewed questions about human nature need to be explored. For example, it now seems crucial to ask how we might better understand human behavior after considering simulated data generated from the study of cellular automata. It also seems crucial to reexamine the importance of the concept of "authenticity" as it relates to fundamental conceptions of truth and the good life.

To a great extent, the success of engaging the "visual turn" will depend upon our capacity to re-think traditional theories and practices of imaging by:

- Revising longstanding views about the relation between "originals" and "copies," between what is "real" and what is "constructed," between "explanation" and "prediction," "reductive" and "emergent" outputs, and between "interpretation" and "translation";

- Generating new conceptions of "representation" and "information";

- Thoroughly reviewing political theories concerning visibility and vision, particularly those ensconced in concerns about "objectification," "reification," "occularcentrism," "surveillance," and "agency."

Since the relevant issues can best be interrogated through interdisciplinary and multi-disciplinary research, I have begun to work through some of these topics by means of collaboration. With Patrick Grim's Group for Logic and Formal Semantics[3], I have been engaged in the construction and analysis of computational models of prejudice reduction. Our primary aim has been

[3] See http://www.computationalphilosophy.org/.

to develop a new account of the mechanisms that can enable the occurrences of prejudice to be reduced. And, with Timothy Engström, I have been working to put together an anthology, *Rethinking Theories and Practices of Imaging*, which provides the basic theoretical vocabulary and crystallizes the basic epistemological, metaphysical, ontological, and social-political issues of the new "visual turn."

Bibliography

Theodor Adorno and Max Horkheimer, *Dialectic of Enlightenment* (Standford University Press, 2002).

Hannah Arendt, *The Human Condition* (University of Chicago Press, 1998).

Hannah Arendt, *Eichmann in Jerusalem: A Report on the Banality of Evil* (Penguin Classics, 1994).

Hubert Dreyfus, *What Computers Still Can't Do* (MIT Press, 1992).

Hubert Dreyfus, *On the Internet* (Routledge, 2001).

Hubert Dreyfus and Stuart Dreyfus, *Mind Over Machine: The Power of Human Intuition and Expertise in the Era of the Computer* (Free Press, 1986).

Andy Clark, *Natural Born Cyborgs: Mind, Technologies, and the Future of Human Intelligence* (Oxford University Press, 2003).

Andy Clark, "Negotiating Embodiment: A Reply to Selinger and Engström." *Janus Head* (forthcoming).

Harry Collins, "Interactional Expertise as a Third Kind of Knowledge." *Phenomenology and the Cognitive Sciences, 3, 2* (2004): 125–143.

Harry Collins, "The Trouble with Madeleine: A Response to E. Selinger and J. Mix." *Phenomenology and the Cognitive Sciences, 3, 2* (2004): 165–70.

Thomas Friedman, *The World is Flat: A Brief History of the Twenty-First Century* (Farrar, Straus, and Giroux: 2006).

Martin Heidegger, *The Question Concerning Technology and Other Essays* (Harper & Row, 1977).

Don Ihde, *Technology and the Lifeworld* (Indiana University Press, 1990).

Don Ihde, *Post-Phenomenology: Essays in the Postmodern Context* (Northwestern University Press: 1993).

Bruno Latour, *Pandora's Hope: Essays on the Reality of Science Studies* (Harvard University Press, 1999).

Herbert Marcuse, *One Dimensional Man: Studies in the Ideology of Advanced Industrial Society* (Beacon Press, 1991).

Evan Selinger and Robert Crease, "Dreyfus on Expertise: The Limits of Phenomenological Analysis." *Continental Philosophy Review 35* (2002): 245–279.

Evan Selinger and Don Ihde, eds., *Chasing Technoscience: Matrix for Materiality* (Indiana University Press, 2003).

Evan Selinger, "The Necessity of Embodiment: The Dreyfus-Collins Debate." *Philosophy Today 47, 3* (2003): 266–279.

Evan Selinger, "Embodying Technoscience." *Journal of Applied Philosophy 20, 1* (2003): 101–107.

Evan Selinger, "Feyerabend's Democratic Argument Against Experts." *Critical Review 15, nos. 3–4* (2003): 359–373.

Evan Selinger, Patrick Grim, et. al, "Reducing Prejudice: A Spatialized Game-Theoretic Model for the Contact Hypothesis." In J. Pollack, M. Bedau, P. Husbands, T. Ikegami, and R. Watson, eds., *Artificial Life IX*. MIT Press, 2004: 244–249.

Evan Selinger and John Mix, "On Interactional Expertise: Pragmatic and Ontological Considerations." *Phenomenology and the Cognitive Sciences 3, 2* (2004): 145–163.

Evan Selinger and Don Ihde, "Merleau-Ponty and Epistemology Engines." *Human Studies: A Journal for Philosophy and the Human Sciences 27, 4* (2004): 361–376.

Evan Selinger, Patrick Grim, et. al, "Modeling Prejudice Reduction." *Public Affairs Quarterly 19, 2* (2005): 95–125.

Evan Selinger and Robert Crease, eds., *Philosophy of Expertise* (Columbia University Press, 2006).

Evan Selinger, ed., *Postphenomenology: A Critical Companion to Ihde* (SUNY Press, 2006).

Evan Selinger, "Normative Technoscience: Reflections on Ihde's Significant Nudging." In E. Selinger, ed., *Postphenomenology: A Critical Companion to Ihde.* SUNY Press, 2006: 89–107.

Evan Selinger, Patrick Grim, et. al, "Game-Theoretic Robustness in Cooperation and Prejudice Reduction: A Graphic Measure." In L. Rocha, L. Yaeger, M. Bedau, D. Floreano, R. Goldstone, and A. Vespignani, eds., *ALife X.* MIT Press, 2006: 445–451.

Evan Selinger and Timothy Engstrom, "On Naturally Embodied Cyborgs: Identities, Metaphors, and Models." *Janus Head* (forthcoming).

Evan Selinger, "Towards a Reflexive Framework for Development: Technology Transfer After the Empirical Turn." *Synthese* (forthcoming).

Evan Selinger, Robb Eason, Robert Rosenberger, Tina Kokalis, and Patrick Grim, "What Kind of Science is Simulation?" *The Journal of Experimental and Theoretical Artificial Intelligence* (forthcoming).

Evan Selinger, "Chess-playing Computers and Embodied Grandmasters: In What Ways Does the Difference Matter?" In B. Hale, ed., *Chess and Philosophy* (Open Court, forthcoming).

Peter Singer and Jim Mason, *The Way We Eat: Why Our Food Choices Matter* (Rodale, 2006).

20

Dan A. Seni

Professor of Innovation and Strategy

Department of the Management of Technology
École des Sciences de la Gestion, Université du Québec à Montréal
Canada

---

1. Why were you initially drawn to philosophical issues concerning technology?

My initial contact with the philosophy of technology came from Ackoff, Jantsch, Ozbekhan, Rapp, Ferré, Mitcham, Agassi, the French historian of technology, Gille, and the philosopher Mario Bunge. My interest in pursuing work in the area was reinforced by the discovery of the philosophy of science of Bunge. As I came to know the man and his work, I took on a second doctoral thesis in the philosophy of science and technology. However, as I reflect on the evolution of my thinking, my interest in the philosophy of science and technology grew as much out of my dissatisfaction with my earlier training as it did from later projects. Perhaps an intellectual biographical sketch may be useful.

In retrospect, I should have gone into science. However, in the 50's and 60's, Canada was a Dominion of the British Empire. The science option was not open to immigrants in the colonies, particularly in the backwaters of Montreal, Quebec. There was little if any science there, then. Indeed, colonialism leaves behind an ethos of marginality that, to this day, permeates the Canadian intellectual landscape.

My early science consisted of the "classics" in physics, chemistry and mathematics, that is, those of the 18th and 19th centuries. We were taught by unenlightened and unmotivated bureaucrats. Science was broached authoritatively with the use of single manuals, government approved, and published in Britain. We were taught laws rather than mechanisms and most of the work was memorization, repetition and exercise. Biology was pre-Darwinian taxonomy and there was not a hint of the idea of the social sciences.

How could we make a living at this, we thought? The bright young lads of the colonies went into engineering, not science.

Scientists and philosophers still today consider engineering to be poor man's science, the family relation with the hand-me-downs of knowledge. No wonder that the philosophy of technology is a subfield of the philosophy of science, in the kitchen rather than in the parlor, so to speak. Consider Popper or Bunge, both important original thinkers in the philosophy of technology, both first and foremost philosophers of science and both of whom dealt with philosophical issues in technology as addenda and appendices to their philosophies of science.

From my education in engineering I came to the view that, generally speaking, engineers, much like their intellectual brethren the accountants, were not interested in ideas. In spite of all this, I came up with a handful of insights that have oriented my work ever since. The first was that it was possible *to use true knowledge, not only technique, to solve real problems.* In other words, I had discovered general or philosophical technology. Second, as an engineer, I began to develop a sense of the way things worked. Indeed, things were complexes or *systems* of other things, and they had *mechanisms* and *functions.* It was clear that if the laws of nature governed the world, events unfolded *mechanismically.* Events or states of affairs were subject to both causal and chance determination. If determination was recursive, it provided a mechanism. And if states of things were mechanismically determined, any particular effect in a causal chain could be an *end*, a *purpose* or a *function* of the determination that preceded it. An end or goal was simply a local equilibrium towards which a system could direct its mechanism. These ideas provided ontological concepts that in turn led to an eventual epistemology for technology.

It was the revolutionary 60's and "everything was possible". So, armed with my insights, I moved on to the social sciences. I became engrossed in applying scientific and technological approaches to the design and engineering of social systems like cities, regions, transportation systems, commercial centers, airports, communities, and even whole economies. My first graduate degree was in economics. Yet, the social sciences were in philosophical turmoil and the areas of sociotechnology positively bubbled over with philosophical problems. I moved on to the University of Pennsylvania where I took a second degree in economics and wrote a doctoral thesis on the planning of social systems.

After a number of years of teaching and doing research, I resigned from the university to practice what I had been teaching. For the next fifteen years I was a professional technologist and entrepreneur. During this time, I had another insight: I found that in the great majority of cases, managers and technologists dealt in commonsense ways with philosophical issues raised in the daily practice of technology. It was clear that *some basis for what we were doing was needed.* I decided to write a second doctoral thesis in the philosophy of science and technology. I am now at the university again where I teach and do research in the management of technology and innovation.

2. What does your work reveal about technology that other academics, citizens, or engineers typically fail to appreciate?

Technology in the broad sense, that is, the ability to anticipate what may happen in the future and to construct an image of different paths, lies at the heart of modern Western culture. In fact, the revolutionary idea of the Enlightenment that defines the boundary between modern times and antiquity is the idea of mastery of uncertainty and risk; the notion that the future is more than the whim or determination of the gods, and that men are not passive before nature. Until a way across this boundary was discovered, the idea of progress was unthinkable and the future was a mirror of the past, the domain of oracles and soothsayers who held dominion over knowledge of events to come.

In general, my philosophy of technology is modernist in the sense of the Enlightenment. The broad goal has been to propose a rationalist realistic philosophical system for technology. Now, there is no body of work in the philosophy of technology comparable to the philosophy of science. Consequently, my approach has been to start with Bunge and continue where he has left off. In that way progress becomes possible and we can avoid the *perennial philosophy.* I view technology pluralistically as a body of knowledge that is neither encompassed by, nor totally contained in science. The problem of identifying and analyzing the characteristics that mark technology off from science has led me to the idea that the core of rational technological action is the idea of *plan.* Hence, the cognitive outputs of technological research and practice, that is, the *rational* dimension of *rational action*, are *plans.* Plans help guide the production and use of artificial states and things. From the perspective of technology, plans are made with the aid of some scientific assumptions and methods. However, technologists also make plans based on local specific *technological theories and methods.*

This view emphasizes a *relative* rational autonomy for technology. It sees technology in terms of thought and reason, not only in terms of artifacts and the social and economic organization of their production and use. Three main ideas underlie it: (1) *Doing* technology, that is, *taking action according to plan* is not the same thing as doing science. Accordingly, the set of problems and methods in technology only partially overlap with those of science. (2) Although related, scientific knowledge and technical knowledge are not of the same kind. (3) Scientific theories provide some, but not the entire basis of technological ones. Philosophy provides some too. Yet there are technological theories as well.

3. What, if any, practical and/or social-political obligations follow from studying technology from a philosophical perspective?

VIEWS OF TECHNOLOGY AND VIEWS OF THE PHILOSOPHY OF TECHNOLOGY

I make a distinction between *the philosophy of technology* and *a philosophy of technology. The philosophy of technology* is a body of literature, "philosophical" commentary, and interpretation that covers a wide range of problems in polemical fashion. It is rarely grounded in scientific theory or in explicit technical practice. On the other hand, *a philosophy of technology* is an ideal: a systematic, consistent body of theses that encompasses ontology, semantics, epistemology, methodology, moral theory and ethics and is in continuity with the philosophy of science, scientific knowledge, evidence and technological practice.

Philosophers in general confuse philosophy of technology with philosophy of science by failing to distinguish between technology and applied science. *Scientists* confuse high technology with basic science. The inter-disciplines of *technology studies* – history of technology, sociology of technology, economics of technology and management of technology - have consistently conflated basic science, applied science, technology, industry and use. This has led to an under-evaluation of the cognitive underpinnings of technological action and an over-emphasis of social and economic conditions. The *manager* conflates it with industry and capital. And the *layman* sees hardware. Finally, *technologists* themselves notably ascribe to the classical separation of thought and action. Consequently, being trained as *technicians*, they are not concerned with philosophical questions. They see knowledge as a source of rules, recipes, norms and procedures, not as the origin of invention or the means of improvement. They prefer know-how to substantive knowledge and are thus vulnerable to received opinion. Coupled with prudence in action, this attitude leads to a highly conservative view of innovation. In sum, the epistemology of practitioners leads to the authority of expertise and to technocracy. Instead, a philosophy of technology ought to provide a basis for action other than self-regulation. In other words, it ought to propose a rational, open, non-authoritative democratic theory of technological method.

Yet all this refers to views about technology itself, not to the philosophy of technology. This is too broad a perspective. So, I will comment briefly on the *philosophy of technology*. For purposes of simplicity, I group points of view in the philosophy of technology into four classes. I personally ascribe to the fourth.

The first is the *ordinary and vulgar view*. On the one hand, orthodoxy in the philosophy of science sees philosophy of technology as simply a working out of philosophy of science. I have argued against this position, proposing that plans and decisions cannot be rationally justified in the full or strong sense that action *necessarily* follows from theory. I conclude that beyond technical rational justification, plans require social, political, moral, and ethical justification. On the other hand, the ordinary philosophy of most technologists leads to a vulgar form of pragmatism. By assuming that (1) scientific propositions can be strongly rationally justified and (2) by undervaluing objective truth in favor of convention, ordinary technology appeals to power rather than reason. A more sophisticated epistemology, both *fallibilist* and *meliorist*, would raise the issue of the basis of justification on rational, on moral and on ethical grounds.

The second position is that of *pragmatism*. In this view, practice precedes and validates theory. Scientific issues arise in attempts to solve practical or technical problems. Science is an extension of technology. Pragmatism clearly understates and undervalues truth in favor of performance. Scientific theories, in this view, are special cases of technological ideas and emerge from social conventions, institutions and disciplines. This idea encroaches on, and brings us to, the third position, namely *sociologism*.

Sociologism and social constructivism see neither science nor technology as rationally autonomous. Both are products of society, not of the mind. They can only be understood as part of social systems. Views that look at technology as essentially social tend to confuse science with applied science and with technology. More important, they fail to address the central question of technology; "How is it done?" It is one thing to ask how technology emerges, how it is organized, and how it can be explained as social process. It is another to ask how to do it. The important and philosophically rich issues emerge when the latter question is raised.

In *neo-rationalism*, *technological theories* form the core of technological knowledge, although they are neither identified nor taught as such. Accordingly, the genesis of technology can be understood as the substitution of pre-scientific technique or craft by science-based knowledge. In Agassi's terms, it is the conversion of magic, craft, tradition and that which is repeatable in action into scientifically justifiable criticizable knowledge.

THE THEORY OF PLANS

A major Platonic error to which Aristotle gave a practical corrective twist, and which exemplifies a materialist theory of mind, concerns the nature and location of ideas. Indeed, whereas for an *idealist* ideas are anterior to the objective world and "live" in some independent realm, for practical and realistic people like Aristotle, ideas are in the mind, or in modern terms, in the brain. In other words, brains think *of* ideas *with* ideas. Consequently, not all ideas are of equal importance; not all of them have the same "thinking power", so to speak. In fact, some ideas are so basic, and are used so often to think of others that we take them for granted. Two basic, but largely unexamined, ideas in technology are those of *system* and *plan*. The former precedes the latter since a plan is, in fact, a system of ideas.

It is, I believe, the task of philosophical analysis and construction to reflect and clarify the basic ideas we use in our thinking. Indeed, the idea of *plan* is so basic and pervasive in technology that its meaning is assumed to be self-evident. Yet, as is common to many basic ideas, self-evidence is a myth.

Basic issues in the philosophy of science deal with the way we represent things of this world in ideas, that is, with the construction of conceptual systems that reflect the world. These are called *scientific theories*. The corresponding issues in the philosophy of technology are also conceptual. However, they refer not to representations but to constructs of action that we make before we act, that is, to constructs which represent our actions and their consequences in ideas of them before we attempt to make them happen. These constructs are called *plans. They are self-forecasts that we attempt to make come true.*

A number of important philosophical problems follow. I have dealt with two of them. (1) What is the conceptual (propositional and logical) structure of a *plan* as a schema for rational action? My approach has been to propose a theory of rational action, in particular, a theory of rational, systemic, collective action in which acting rationally is to act according to plan. (2) What arguments respond to the practical necessity of rationally justifying a plan? My approach here has been to propose a theory of partial, progressive, rational justification in which a plan is justified if it is rationally superior to its alternatives on the basis of the best available knowledge under practical contingencies.

Three related ideas lie at the root of a rational theory of technology. The first is, as I have said, that the core of technology consists of criticizable concepts and propositions organized in *plans.* Further: A *plan* is a conceptual system or construct explaining rational action in general and only as corollary in technology. Thus, the concept of plan is more general than its use in technology.

The second is that of an *active* social organization or *sociotechnical system,* a supra-individual agency acting on the basis of *sociocognitivity.* Of course, technologies emerge, develop, evolve and are used in the context of social relationships. In a trite sense then, all technologies are social. Once accepted however, this idea leads to the distinction between sociotechnical systems or *active social systems* that produce things, and reactive or passive social systems that are either the objects of action or emerge as the unplanned consequences of other social processes. In short, I propose a theory of social systems that incorporates a theory of plans. In fact, only *active organizations* use technological theories to make plans.

The third is the idea of technological method as distinct from scientific method. Technological method follows from realistic and practical assumptions about "doing" technology in terms of knowledge-based action including learning, research, planning and communicating in social systems. The method-as-process view of the lone researcher in science consistently confuses epistemology in science with epistemology in technology. Indeed, by confounding the relation of subject-object with agent-patient, the *spontaneous philosophy of technology* has assumed a one-body model of action.

A realistic technology involves a two-body model of action and three ontic components, an agent, a patient and a helper. And thus, rather than a single subject-object relationship, there are two; an *internalist* view (the agent) and an *externalist* view (the observer of the agent-patient dyad). "Doing" technology requires that the latter improve the former. This has led to a formulation of technological method as a *method of delegated agency* (my neologism) in terms of helping another in attempting to achieve an end.

As regards epistemology, the theory of plans is a rational action response to irrational philosophies of technology of sociologistic, economistic, phenomenological, semiotic or spontaneous kinds. Indeed, I use the same concept of human action that first came to us from Socrates and Aristotle: We act on knowledge of our beliefs and desires. Action thus combines matters of fact with matters of knowing. If improvement in knowledge improves belief and desire, it improves action. This relation between reason and action links an act both to its consequences and to the way an agent knows the act to be part of a plan or project. Of course, this idea is simply an application to action theory of the methodological principles of explanation in social science proposed by thinkers as diverse as Max Weber, Popper, Habermas, Winch, Jarvie and Bunge. The claim to rationality in action is a claim of competency based on claims of validity in knowledge. Moreover, a rational technological epistemology claims that *partially* true but *improvable* knowledge is all that is required for improving the effectiveness of action.

What is technological knowledge in the sense above? The core of knowledge in a technical discipline or profession is the class of available technological theories. At first glance, technological theories come in two varieties; (1) *substantive theories*, that is, theories directly about artifacts and (2) *operative theories* about the action of agents on objects, that is, about agent-object supersystems. *Substantive technological theories* are really extensions of scientific theories to artifacts. *Operative theories* relate the action and decisions of agents to the artifact. Decision theory, design theory, systems theory, control theory, computing theory, management theory, planning theory are examples of *general operative technological theories*. Thus, operative technological theories refer objectively to agent-object interaction. Operative technological theories may either be used descriptively e.g. decision theory, or, given appropriate pragmatic transformations, prescriptively, e.g. a theory of optimal choice.

Substantive technological theories are entailed by scientific theories plus additional assumptions and data. On the other hand, operative theories are specifically technological, and may rest on little or no explicit science. Many technologies tackle problems without much substantive theory (e.g. urban planning, economic development, environmental design, and software engineering) but there are none that don't have some operative ones. Some technological theories are science-based and science-driven (e.g. medicine, biotechnology and engineering); others are not (e.g. information and communication technology, computer science, operations research, law and architecture.) Finally, some are very broad (e.g. the general theory of structures) while others are narrow, local or specific (e.g. the plan for a given structure) Plans are the lowest level of technological theory and lead directly to rule-based action.

4. If the history of ideas were to be narrated in such a way as to emphasize technological issues, how would that narrative differ from traditional accounts?

The history of artifacts is both rich and vast. And since the artifact cannot be completely dissociated from ideas predominating at the time of its invention or production, there is, in the history of the artifact, an implicit sub-narrative of the evolution of technological ideas. There is not, however, a distinct history of technological ideas that parallels the history of science since there is as yet no theory of technological theories. If technology as knowledge and technological theory come to be articulated, then the history of technological ideas may one day be written. Here is a sketch of some *desiderata* for such a history.

It would, of course, be rooted explicitly in a systematic philosophy of technology. It would emphasize the analysis of ideas specific to a social system or culture, and relate them to material conditions of the time and place. This would reduce the emphasis in the history, sociology and economics of technology on the social conditions associated with technological culture. History would bear on ideas rather than on social arrangements, as it does say in the work of Lewis Mumford and the sociology of technology. This would have the advantage of documenting "local" technological cultures within "micro" social systems such as large firms, organizations and bureaucracies. We could, for example, have the history of jet engine technology at GE, or the history of World Bank econometric modeling technology.

It would be based on a theory of history or a "dynamics" of technological theories much as Kuhn and Popper have written for the origin and the history of scientific theories. The concept of "technological paradigms" could be documented rather than be pronounced.

It would describe both the interaction and the relation between the history of scientific ideas and the history of technological ones, as well as the commonality of both (see e.g. Norbert Wiener, *The Care and Feeding of Ideas*) It would propose a history of the influence of technological ideas on the progress of science, particularly in the biological and medical sciences.

It would underscore the influence of systems of ideas and world-views and their evolution on the history of technological thought and material conditions. There is, for example, no work that links the intellectual history of the Enlightenment with the birth of western technological ideas. Neither is there a historical analysis of "origins", of why technology flourishes at certain times and in certain places and stagnates in others.

Finally, economic theory, in general, has considered technology to be "exogenous" to the economic system. Modern economic theory has just only begun to deal seriously with technology. Thus, economic history could to rewritten as socio-technological-economic theory.

5. With respect to present and future inquiry, how can the most important philosophical problems concerning technology be identified and explored?

The issue is not how inquiry in philosophy ought to proceed: Obviously, the only way to identify important philosophical problems is through criticism and construction. And clearly in order to promote this kind of inquiry it is necessary to train technologists themselves, *rather than other philosophers*, in the philosophy of technology as well as in contemporary issues raised by scientific and humanistic world-views. But the issue seems to me to be rather *what* are the most important questions for the philosophy of technology to tackle.

I have already outlined what I consider to be the substance of a full and systematic treatment of *a philosophy of technology*. The *social issues that the philosophy of technology* must deal with are another matter. The French philosopher, Michel Serres, in the inimical style of French literary philosophy, has addressed a number of them raised by advanced technology. He has dealt at length with two of them, namely: (1) the idea of a "natural contract" between men and the shrinking disappearing natural world on earth, and (2) the rise of a rational "global techno-humanity" with its attendant individualism and loss of culture and community. Three other issues seem to me to be important; (1) the natural–artificial dichotomy and the implications for sociotechnical systems, for human settlement of the earth, for biological health, and for the very definition of what it is to be a human being rather than a cyborg; (2) the growth and systematic development of long term technological forecasting and assessment and (3) the construction of serious realistic sociotechnological utopias, work on which came to a halt in the early twentieth century with the rise of totalitarianism.

Finally, a mature philosophy of technology needs to respond to a number of *philosophical issues.* These provide a broad research programme. Here are suggestions for some: (1) A theory of pragmatic partial truth and a new theory of risk; (2) A modern theory of rational (knowledge-based) action; (3) A materialist, realist systemic ontology of the artifact. This kind of investigation may reignite work in systems theory because technologists work on real objects, and realism calls for systemism. Yet, though the "systems approach" has been professedly adopted in systems engineering and in general engineering, it has yet to be integrated in technology as it has, say, in biology. (4) A modern theory of value, in particular a theory of practical or technological value; (5) A theory of function and artifactual function modeling and explanation as a prelude to a theory of design; (6) A philosophical theory of work and production; (7) There is a definite need for an epistemology of technology and its subfields in logic and semantics. The epistemology of technology must account for the fact that technological knowledge is often local or idiosyncratic to an organization, and rather than be explicit may be "embedded" in various ways. Moreover, it must also account for the fact that agents are rarely bare individuals; rather, they are social systems, organizations and institutions. It must therefore account for the idea of "organizational learning" or sociocognitivity. (8) Finally, there is a need for basic work on the methodology of technology. R&D is applied science, not technology, and therefore it uses scientific method. Unlike applied science, technological research always involves design and always pursues the interest of an agent. The technologist is a hired hand, paid to serve another.

Bibliographical Note

The following is a short list of some non-technical publications on these ideas: On a succinct statement of the theory of plans see “The socio-technology of sociotechnical systems; elements of a theory of plans” in Weingartner and Dorn (Eds.), *Studies on Bunge's Treatise in Basic Philosophy*, Rodopi, Amsterdam, pp. 431–454, 1989. On the distinction between planning theory and the theory of plans see “Planning theory or the theory of plans?” in Kuklinski, A. (Ed.) *The Production of Knowledge and the Dignity of Science*, Rewasz, Warsaw, Poland, pp. 131–146, 1995. On sociotechnical systems see “Planning as sociotechnology” in Kuklinski, A. (Ed.) *The Production of Knowledge and the Dignity of Science*, Rewasz, Warsaw, Poland, pp.147–159, 1995. On the concept of technical value and the theory of work see “From scientific management to process engineering: The spontaneous theory of technological value in the design of work in organizations”, *Journal of Construction Research, Vol.2*, pp.91–98, 2000. On future modeling as an approach to design and invention based on system functions see “Function Models: A General Framework for Technological Design” *Value World, Journal of the International Society of American Value Engineers, Volume 28, No.2*, pp.8–11, 2005.

On more technical development of the theory of plans see “Elements of a Theory of Plans”, Doctoral dissertation, University of Pennsylvania, Philadelphia, Pa., 1994.

Finally, I would be happy to respond to requests for further material, working papers and work in progress.

21

Peter Singer

Ira W. DeCamp Professor of Bioethics

Princeton University
USA

---

1. Why were you initially drawn to philosophical issues concerning technology?

My field is ethics. Ethical judgments tend to be highly resistant to change. Technology, on the other hand, changes rapidly. I'm interested in the way that traditional moral views need to be modified in order to apply to the very different world that technology has brought about. Let me give you some examples:

- In the 1960s, doctors began using respirators to maintain bodily functions in patients with severe brain injuries – and in some cases, no brain function at all. Around the same time, it became possible to transplant organs – initially kidneys, then hearts – into patients with diseased organs. The combination of these two advances led to a change in the definition of death, so that patients with no brain function could be declared dead, cease to take up beds in intensive care units, and be used as organ donors while their organs were still in good condition. Now that we have the technology to take images of the brain, and see which parts are functioning, we could move still further, and declare dead those who have no functioning cortex, and hence will never recover consciousness. Should we do that?

- In 1978 Robert Edwards and Patrick Steptoe succeeded in fertilizing a human egg in vitro and returning the resulting embryo to the uterus of the woman from whom the egg came. The embryo implanted and grew into a normal child. So suddenly we had to face a new ethical issue: what is the moral status of an embryo outside the human body? Many other ethical issues have flowed from that initial success, including issues of embryo donation, research on embryos, genetic selection prior to implantation of the embryo, and cloning.

- Not all that long ago, we could realistically only be responsible for the poor in our village, or at least our region. Over the last century, we have developed methods of communication that enable us to know more or less immediately when a drought or a flood or other natural catastrophe causes people to be in a life-threatening situation. And we have also developed the technology to bring life-saving aid to them. What obligations does this change bring? How much should citizens of the developed nations be doing for those living in life-threatening poverty in other parts of the world?

- One last example. Again, not all that long ago, the animals we raised went out and gathered things we could not or would not eat. Cows ate grass, chickens pecked at worms or seeds. Now the animals are brought together and we grow food for them. We use synthetic fertilizers and oil-powered tractors to grow corn or soybeans. Then we truck it to the animals so they can eat it. When we feed grains and soybeans to animals we lose most of their nutritional value. The animals use it to keep their bodies warm and develop bones and other body parts that we cannot eat. Pig farms use six pounds of grain for every pound of boneless meat we get from them. For cattle in feedlots, the ratio is 13:1. Even for chickens, the least inefficient factory-farmed meat, the ratio is 3:1. Factory farmining is not sustainable. It is also the biggest system of cruelty to animals ever devised. In the United States alone, every year nearly ten billion animals live out their entire lives confined indoors, often dependent on antibiotics to keep them going. Hens are jammed into wire cages, five or six of them in a space that would be too small for even one hen to be able to spread her wings. Twenty thousand chickens are raised in a single shed, completely covering its floor. Pregnant sows are kept in crates too narrow for them to turn around, and too small for them to walk a few steps. Veal calves are similarly confined, and deliberately kept anemic. How does this system of raising animals affect the ethics of eating meat and other animal products?[1]

2. What does your work reveal about technology that other academics, citizens, or engineers typically fail to appreciate?

I'll focus on ordinary citizens, rather than academics or engineers, in answering this question. As I said in response to the previous question, ethical judgments tend to be resistant to change.

Therefore our judgments often do not handle technological change very well.

But there are interesting differences in the various areas I mentioned above, and it isn't always clear why. Take new reproductive technology. Most people think that it is a good thing if couples who want to have a child are able to have one. Religious organizations, in particular, regard marriage as something to be cherished, and they see its functon as procreation. So you would think that when a new technology develops that makes it possible for infertile couples to have children, they would approve. But many of them do not. The Roman Catholic Church, for example, focuses on the fact that the child was not produced by sexual intercourse, and rejects *in vitro* fertilization.

On the other hand, in the case of the definition of death, the same religious bodies effectively handed the decision over to the medical profession. This is curious, because it is not as if there were any new medical or scientific discovery made regarding death. All that happened was that doctors developed criteria that showed that certain patients, whose brain function had ceased, would never recover. But their hearts were still beating, their skin remained pliable, their bodies warm to the touch. In effect the doctors were saying: "Here are some human beings who are, now, considered alive. They are very severely injured and will never recover. We think it would be better if they were considered dead, so that we could cut them open, and give their organs to strangers." And the churches and so-called "pro-life" groups said "OK." Isn't that amazing? In this case it seems that the life-saving benefits of the organ transplants, and the futility of using advanced medical technology to keep alive patients who would never recover consciousness, were just too clear for them to resist. But of course

[1] For details see Peter Singer and Jim Mason, *The Way We Eat,* Rodale, New York, 2006.

they didn't say that. Instead they rationalized the decision in a variety of ways that, as I have shown in detail in a book called *Rethinking Life and Death,* simply do not hold up.[2]

The technology used to produce food tends to be ignored. People focus on the product – the pork chop, the eggs – and see it as the same thing, even if the pigs and hens have lived very different lives and the implications for environmental sustainability have changed drastically. That attitude needs to be changed.

3. What, if any, practical and/or social-political obligations follow from studying technology from a philosophical perspective?

This is exactly what my work focuses on. The implications are vast. Thomas Aquinas has been the dominant philosophical influence on the Roman Catholic Church for the past five centuries. He's normally thought of as a very conservative figure. Yet look at this quotation from Aquinas: Whatever a man has in superabundance is owed, of natural right, to the poor for their sustenance. So Ambrosius says, and it is also to be found in the *Decretum Gratiani*: "The bread which you withhold belongs to the hungry: the clothing you shut away, to the naked: and the money you bury in the earth is the redemption and freedom of the penniless."[3]

In Aquinas's time – the thirteenth century – the Church helped the poor of each parish by collecting a tithe from the rich – ten percent of their income was supposed to go to support the poor. But now that technology has extended the reach of our ability to help the poor, even the Church fails to take seriously the views of its own greatest philosopher. Yet the gap between the rich citizens of affluent nations and the 1.2 billion people living below the poverty line set by the World Bank is so vast that it would take much less than a tenth of the income of the rich to eliminate most of the poverty in the world. Jeffrey Sachs, a leading economist, has estimated that it might take as little as 0.6% – just 60 cents in every $100 we earn.[4] Maybe it will be twice or even five times that amount, but even that would only be 3% of our income. I would argue that our obligations extend *at least* to that.

---

[2] Peter Singer, *Rethinking Life and Death,* Text, Melbourne, 1995, St Martin's Press, New York, 1996.

[3] Thomas Aquinas, *Summa Theologica,* II-II, Q 66 A 7.

[4] Jeffrey Sachs, *The End of Poverty,* Penguin Press, New York, 2005.

I also think we have an obligation to be more conscious of the difference that technology makes to the ethics of what we eat, both in regard to the treatment of animals, and in regard to sustainability. For instance, there has been a strong recent move to eat food that is locally produced, rather than imported long distances. That is itself a response to the technology of cheap transport, usually by road, which means that the average item on an American's plate has traveled about 1500 miles to get there. That has environmental costs, of course, because it takes a lot of fossil fuel to move it those distances, and that means a lot of greenhouse gas emissions. So eating locally is normally a good thing. But not always. If you live in Connecticut and eat locally-produced tomatoes in June, they will have been grown in a greenhouse, and the oil used to heat the greenhouse could have been used to truck them up from Florida. Similarly, because rice grown in California uses a lot of energy, and transport by ship is very energy-efficient, compared to road and air transport, Californians may do better for the planet if they buy rice from Bangladesh rather than rice grown in their own state.[5]

4. If the history of ideas were to be narrated in such a way as to emphasize technological issues, how would that narrative differ from traditional accounts?

I think it would shed a lot of light on the processes of change. We would need to consider in detail the interaction of technology and other ideas, especially our values. Of course, this has been attempted, especially by marxist historians. Often marxist accounts of history are too rigid, or deterministic. But the underlying idea, that the technology we have for producing our food and meeting our other needs and desires has a huge impact on our ideas and values, is surely correct. Marx would not have been at all surprised to see that, in an age of instant communication and cheap transport, we have moved towards a more global economy, and that this should have political and social consequences. He would also understand that, despite all the rhetoric we hear about the value of the independent family farm and the rural lifestyle, once big corporations develop ways of producing meat and eggs more cheaply than family farmers are able to do, the family farmer rapidly becomes an endangered species, and small rural towns become ghost

[5] For details, see *The Way We Eat.*

towns – while neither government nor consumers do anything to stop farms being turned into vast factories for producing animal products.

As I've already indicated, it would be epecially interesting to consider why we are so accepting of some technological change, and so resistant to others. What is it about cloning, for example, that produces such widespread opposition, when factory farming does not?

5. With respect to present and future inquiry, how can the most important philosophical problems concerning technology be identified and explored?

It will be apparent from what I have already said that I think the most important philosophical issues concerning technology are ethical ones: how do these changes affect our values, and hence, our actions? I've already referred to the effect of new methods of farming, and of food transport, on fossil fuel use. Greenhouse gas emissions are a grave threat to the world as we know it. In the next few decades, hundreds of millions of people could become climate change refugees because changes in rainfall, or rising sea levels, have deprived them of the means of growing their own food. The rich nations are overwhelmingly responsible for this. In historical terms, they are the ones that have caused the problem, by their heavy use of fossil fuels. And at present-day rates of emission, too, the United States is still putting into the atmosphere about five times as much carbon, per person, as China, about 15 times as much as India, and 200 times as much as Ethiopia. This is fundamentally a question of justice. The capacity of the atmosphere to absorb our waste gases, without causing unpredictable and very probably disastrous climate change, is strictly limited, and the developed nations have grabbed the biggest slice of it, leaving little or none for other countries.[6] Just as we need to think about our responsibilities to the poor in terms of a global, not just a national ethic, so too we need to consider our responsibility for our waste gases from a global perspective.

[6] For details, see my *One World,* Yale University Press, New Haven, 2002.

Selected Bibliography

Democracy and Disobedience, Clarendon Press, Oxford, 1973; Oxford University Press, New York, 1974; Gregg Revivals, Aldershot, Hampshire, 1994.

Animal Liberation: A New Ethics for our Treatment of Animals, New York Review/Random House, New York, 1975; Cape, London, 1976; Avon, New York, 1977; Paladin, London, 1977; Thorsons, London, 1983.

Animal Rights and Human Obligations: An Anthology (co-editor with Thomas Regan), Prentice-Hall, New Jersey, 1976. 2nd revised edition, Prentice-Hall, New Jersey, 1989.

Practical Ethics, Cambridge University Press, Cambridge, 1979; second edition, 1993.

Marx, Oxford University Press, Oxford, 1980; Hill & Wang, New York, 1980; reissued as *Marx: A Very Short Introduction*, Oxford University Press, 2000; also included in full in K. Thomas (ed.), *Great Political Thinkers: Machiavelli, Hobbes, Mill and Marx*, Oxford University Press, Oxford, 1992.

Animal Factories (co-author with James Mason), Crown, New York, 1980

The Expanding Circle: Ethics and Sociobiology, Farrar, Straus and Giroux, New York, 1981; Oxford University Press, Oxford, 1981; New American Library, New York, 1982.

Hegel, Oxford University Press, Oxford and New York, 1982; reissued as *Hegel: A Very Short Introduction*, Oxford University Press, 2001; also included in full in German Philosophers: Kant, Hegel, Schopenhauer, Nietzsche, Oxford University Press, Oxford, 1997.

Test-Tube Babies: a guide to moral questions, present techniques, and future possibilities (co-edited with William Walters), Oxford University Press, Melbourne, 1982.

The Reproduction Revolution: New Ways of Making Babies (co-author with Deane Wells), Oxford University Press, Oxford, 1984. revised American edition, *Making Babies*, Scribner's New York, 1985.

Should the Baby Live? The Problem of Handicapped Infants (co-author with Helga Kuhse), Oxford University Press, Oxford, 1985; Oxford University Press, New York, 1986; Gregg Revivals, Aldershot, Hampshire, 1994.

In Defence of Animals (ed.), Blackwells, Oxford, 1985; Harper & Row, New York, 1986.

Ethical and Legal Issues in Guardianship Options for Intellectually Disadvantaged People (co-author with Terry Carney), Human Rights Commission Monograph Series, no. 2, Australian Government Publishing Service, Canberra, 1986.

Applied Ethics (ed.), Oxford University Press, Oxford, 1986.

Animal Liberation: A Graphic Guide (co-author with Lori Gruen), Camden Press, London, 1987.

Embryo Experimentation (co-editor with Helga Kuhse, Stephen Buckle, Karen Dawson and Pascal Kasimba), Cambridge University Press, Cambridge, 1990; paperback edition, updated, 1993.

A Companion to Ethics (ed.), Basil Blackwell, Oxford, 1991; paperback edition, 1993.

Save the Animals! (Australian edition, co-author with Barbara Dover and Ingrid Newkirk), Collins Angus & Robertson, North Ryde, NSW, 1991.

The Great Ape Project: Equality Beyond Humanity (co-editor with Paola Cavalieri), Fourth Estate, London, 1993; hardback, St Martin's Press, New York, 1994; paperback, St Martin's Press, New York, 1995.

How Are We to Live? Ethics in an Age of Self-interest, Text Publishing, Melbourne, 1993; Mandarin, London, 1995; Prometheus, Buffalo, NY, 1995; Oxford University Press, Oxford, 1997.

Ethics (ed.), Oxford University Press, Oxford, 1994.

Individuals, Humans and Persons: Questions of Life and Death (co-author with Helga Kuhse), Academia Verlag, Sankt Augustin, Germany, 1994.

Rethinking Life and Death: The Collapse of Our Traditional Ethics, Text Publishing, Melbourne, 1994; St Martin's Press, New York, 1995; Oxford University Press, Oxford, 1995.

The Greens (co-author with Bob Brown), Text Publishing, Melbourne, 1996.

The Allocation of Health Care Resources: An Ethical Evaluation of the "QALY" Approach (co-author with John McKie, Jeff Richardson and Helga Kuhse), Ashgate/Dartmouth, Aldershot, 1998.

A Companion to Bioethics (co-editor with Helga Kuhse), Blackwell, Oxford, 1998.

Ethics into Action: Henry Spira and the Animal Rights Movement, Rowman and Littlefield, Lanham, Maryland, 1998; Melbourne University Press, Melbourne, 1999.

Bioethics. An Anthology (co-editor with Helga Kuhse), Blackwell, 1999/ Oxford, 2006.

A Darwinian Left, Weidenfeld and Nicolson, London, 1999; Yale University Press, New Haven, 2000.

Writings on an Ethical Life, Ecco, New York, 2000; Fourth Estate, London, 2001.

Unsanctifying Human Life: Essays on Ethics (edited by Helga Kuhse), Blackwell, Oxford, 2001.

One World: Ethics and Globalization, Yale University Press, New Haven, 2002; Text Publishing, Melbourne, 2002; 2nd edition, pb, Yale University Press, 2004; Oxford Longman, Hyderabad, 2004.

Pushing Time Away: My Grandfather and the Tragedy of Jewish Vienna, Ecco Press, New York, 2003; HarperCollins Australia, Melbourne, 2003; Granta, London, 2004.

The President of Good and Evil: The Ethics of George W. Bush, Dutton, New York, 2004; Granta, London, 2004; Text, Melbourne, 2004.

How Ethical is Australia? An Examination of Australia's Record as a Global Citizen (with Tom Gregg), Black Inc, Melbourne, 2004.

The Moral of the Story: An Anthology of Ethics Through Literature (co-edited with Renata Singer), Blackwell, Oxford, 2005.

In Defense of Animals. The Second Wave (ed.), Blackwell, Oxford, 2005.

The Way We Eat: Why Our Food Choices Matter, Rodale, New York, (co-author with Jim Mason); Text, Melbourne; Random House, London, 2006.

22

Susan Leigh Star

Professor, Women and Gender Studies, and Senior Scholar, Center for Science Technology and Society

Santa Clara University, California
USA

---

1. Why were you initially drawn to philosophical issues concerning technology?

Growing up in rural, working-class, Rhode Island, I was of course surrounded by technology. Cars. Tractors. Do-It-Yourself Everything in our home. Bicycles. Sewing machines. Ovens. Gardens. Worms to sell for bait. Buckets for the backstairs bucket-brigade to get rid of water in the basement during heavy rains. A small television. A typewriter for my 11th or 12th birthday. A few books. Lawn mowers. Knitting needles. Hunting and fishing gear. A shotgun. A freezer.

My father worked with technologies all the time, first as a painter/wallpaperer, and our house was beautiful as a result. Later, the small downtown of the closest city died, the paint and wallpaper store site. The store closed and my father became a salesperson for parts of things in factories: grinding wheels, drill bits, and finely honed glass pieces. My mother worked initially as a bookkeeper and clerk at a local oil provider, then, for a very long time, as a telephone receptionist at a wholesale grocer's headquarters, IGA. Again, I would have said, if asked, that no, they really didn't have anything to do with technology. Except one day I visited my mother after school and found her at her desk, talking into the air. She had on an invisible neck-circling headphone! I think I would have called that technology.

As a child, there were many ways that we learned technology in my family. When we got to play in the paint store, we would run the brushes over our faces, feeling the relative softness of brushes numbered 2, 3, 4. And the materials with which they were made, increasingly synthetic, but my grandfather always kept some of the "good stuff" in a glass case. Once the famous clown, Emmett Kelly, Jr. came in and did just what we did, run brushes over his face, and finally bought one of our good ones. That story circled for years at various dinner tables, at one relatives' or another. We would also take the little color chips and learn to "read" them, expanding them until we could envision a .5inch line of color as a whole house or room. Similarly, with discarded wallpaper sample books, we would wallpaper our dolls' houses, write notes to each other, and envision our grown-up houses.

My mother often brought home the tryout samples that IGA would give to its stores: chocolate breakfast cereal, pink and blue cheerios, canned pasta, and bonuses from the moon launch: peanut butter sticks and Tang (feh) which were spin-offs from the race to space. We ate them all. I shudder to think.

The most curious technoscience adaptation (as I would now call it) was the following: my mother was in charge of a switchboard with perhaps 20 lines. At busy times, she would answer the phone "Roger Williams Foods," and begin a conversation with me. This conversation would be punctuated every 10-120 seconds with the need for her to answer another line, say, "Roger Williams Foods, may I help you," plug in the customer, and then continue the conversation as if the interruption never happened. I so naturalized this way of talking that I didn't even remember it until years later when I began thinking about everyday technology, and women's work as telephone operators.

If anyone had asked me then if I interacted with technology, I would have stared, thought about it, and said, absolutely not. Except maybe TV. I associated "technologies," so named, as being about Big Science, or perhaps Big Engineering. I didn't know anyone who did that. Unconsciously, tools and humble technologies of the sorts around me seemed like heavy anchors to the working class, a class I seemed to have been born struggling to leave. Having a lifetime job as a beautician or a factory worker was my nightmare. My closest uncle worked for 30 years on the night shift at the "new, automated" post office in Providence. Our school classes were taken on field trips to the facility; these trips were pretty acutely boring. Uncle Tommy slept all day, and we had to be careful not to call, yell, or wake him around the house or in the yard. Even after retirement, he was permanently altered to be a night owl, watching movies until dawn and then sleeping late. Yet this was a good job, a union job, increasingly rare during the

1960s and 70s, when all around us layoffs became temporary jobs at places like MacDonald's.

The only technical stuff that seemed friendly, approachable, and a door-opener, not a ball-and-chain, was print technology (again, not my word at the time!). I had started to read freakishly early at the age of two, and the words on the page beckoned me to ways of life I yearned for. Even if I rarely knew how to pronounce the arcane words, I longed to talk about the issues raised in the books. I literally read dictionaries. Let me say it again: This was not technology. Girls didn't do *that*, especially not rural, working-class girls. The small moral panic about getting more bright girls in engineering, or mathematics, was decades away. If I had been presented with science then, I suspect I would have liked it. But while we had a few courses in chemistry, biology and physics in high school (no shop, of course, just the really really simple (sic) things like baking a loaf of bread or making a dress in home economics), there was no imaginary in my life that would connect that class work to making a living, not from there to travel, conversations with interesting people, reading, reading, and not from there to writing and talking about ideas.

My lived experience was that everyday unmarked dull, half inhabited traps, firmly anchored in a working class (two words that also meant nothing to me until graduate school) lifeworld. For girls, that lifeworld meant husband, children, housework, and work of the kind all my female relatives and friends undertook: the beauty parlor, the telephone company, bedpans in hospitals as a nurse's aid, caring for disabled people in their homes or in a nursing home, becoming a cashier at a local supermarket. I was in dread of what I saw before me, in this sense – I thought that people in this place worked fifty weeks a year, "lived" for the weekends and the two-week vacation somewhere even more rural than where we lived, and never got to talk about anything except local sales on clothing or food, babies, boyfriends/husbands, the weather, and sometimes, television.

I was, in short, starving for philosophy – for friends to talk with about how to grow up and be educated, how to ask questions without being ridiculed, how not to be a *GIRL*. My best friend in high school was an ex-nun, and she agreed to talk with me and teach me a bit about theology. I began to read thinkers like Teilhard de Chardin, Harvey Cox, St. Augustine – whatever semi-random items our tiny high school library had to offer. This helped me begin to breathe and kick my way into the water.

I was quite serious in my quest for these conversations. I obtained a scholarship to Radcliffe College, and began taking classes in philosophy (beginning with the requisite symbolic logic). I was told that I couldn't major in religion, since, after all, that was a "vocational degree." (Although I had stated this intention on my application.) I didn't really have a starting place, except for hunger, and once again I began to despair of finding people to really talk with. During this phase, I was also extremely shy and aware of my homemade clothes, my ignorance of practically everything except what I had read. When one of my roommates said casually that she was going skiing in Gstaad over Thanksgiving, and asked where I would be heading, I muttered something vague about staying around for the peace and quite. I could not say the truth then, that I would be taking the Greyhound Bus to Providence.

So, for all these reasons, after freshman year, I dropped out, married a lovely young hippie, became a serious meditator, and went to Venezuela to co-found an organic commune. Robert (Harvard '70) and I talked non-stop about everything I longed to know. I read books on farming, and on communes, and on the technologies of self – meditation, psychedelic drugs, social movements. We carried truckloads of books, as well as our wedding crystal, via Volkswagen and then mule to our home, which was about 2.5 hours walk from the end of the bus route. During this time, the women's movement became an explosion in the US, and somehow, on a visit to Caracas, I found a copy in English of Kate Millett's *Sexual Politics* (1978 [1970]). Reading by kerosene lamp at our farm deep in the Andes, I somehow came up with the question that has gripped me to this day: how does good technology ((e.g.) organic farming,) connect with changing one's self? How does bad technology (e.g. clitoradectomy, guns) connect with the larger structure of the world? And vice-versa. Recognizing these now as philosophical (or perhaps spiritual) questions, I read all sorts of things when chances to buy or trade books came along: Buckminster Fuller, theology, Ruth Stout on how to compost. Baba Ram Dass. Later, when I returned to the States and went back to college, I added books on rape, orgasm, evolution, the brain, community organizing, consensus reality; books by Gregory Bateson, feminist theory, Chogyam Trungpa Rinpoche, Alan Watts, D. T. Suzuki, William James and a thousand mimeographed newsletters on permaculture, pesticides, solar panels. I learned how to order organic oranges by the crate directly from Florida, and how to drive to the cargo port at Logan Airport in freezing sleet to pick them up in our 1951 Chevrolet truck. Still, I was unaware that there might be a group that studied these sorts of questions professionally (by now I had heard of philosophers, but had gotten bogged down in the secret language they seemed to be in on, and that I never could seem to grasp). Philosophy of technology as a formal field, then, eluded me until graduate school. At the same time, these earlier experiences formed a powerful root network; now the question was, how could these pieces possibly fit together?

2. What does your work reveal about technology that other academics, citizens, or engineers typically fail to appreciate?

I have been trying, for about the last twenty years, to link the following three dimensions: lived experience, technologies (both everyday and those at some remove, such as linear accelerators and nuclear power plants), and silences. This has meant, ironically, returning to the stuff of my childhood and re-seeing many aspects of it as technology, and then to face the plain, terrible consequences of why I couldn't see it that way in the first place. I had eventually succeeded in my goal of moving away from Rhode Island, and of leaving the working class, if indeed one ever really leaves one's class of birth. I Pygmalioned myself into becoming an academic, after it seemed clear that organic farming in Venezuela was a self-limiting choice. I found many different streams of action around which and within which I wove a new self. And for the past several years, I have been tying to image a way of telling about technology that makes it possible to see my origins as nuanced, valuable, and open to change.

So the inclusion of everyday items, of all forms of technology, and its contextual nature, are things that I bring to the analysis of technology. Technology is a living extension of our bodies, a blunt mediator of our tongues, a doorway or a stone wall, depending on where we stand. I brought the concept of boundary objects into the world of science studies, and the nature of these objects attests to the same interpretive multi-valency, and the same obduracy.

3. What, if any, practical and/or social-political obligations follow from studying technology from a philosophical perspective?

In college several of my teachers were methodological skeptics, including Robert Rosenthal (who did pioneering work on anomalies in psychology) and Mary Daly (a true iconoclast, who railed against "methodolotry."). As a psychology major, I learned to apply electrodes, read printouts, be wary of lie detectors (and WHY). At the same time, I read about experimenter expectation effects, and began to turn around and think – HOW then to know what another is thinking? Do technologies, all technologies, simply get in the way of perception? These questions fascinated me.

So when the opportunity came, in graduate school, to get to know science and technology more intimately, I wasn't just drawn to philosophy of science/technology, I flew to it. I found philosophical questions embedded in conversations and ideas. I found them in social movements (eco-feminism, feminist philosophy of science, feminist science studies, biology and gender). So I began philosophy of technology with a struggle to read technologies such as EEGs as connected to consciousness, thought, or language, and to see methodological anomalies as political.

The obligations that result from this way of perceiving frame my whole life. I am obligated *to* philosopher friends – John Dewey, William Wimsatt, James Griesemer, Arthur Bentley – for making such a wide space for asking the most serious questions possible. Whose knowledge counts? And why? Are there patterns that flow across our naturalized nature/culture boundary, recursing and repeating themselves, and how shall we teach each other to read them? I am obligated to a tenacious honesty in trying to speak about what I find, and to write about it clearly. This obligation is sometimes onerous, especially when my fingers are slow and my tongue is thick; at other times, especially when co-building our imaginary in conversation, it is featherlight, giddy, redolent with the best parts of teaching.

4. If the history of ideas were to be narrated in such a way as to emphasize technological issues, how would that narrative differ from traditional accounts?

It would include mess, suffering, exclusion, manual labor, and invisible work. I have tried to do this in all the historical work cited below. And, as the poet Muriel Rukeyser said in the poem, *Käthe Kollwitz*: "What would happen if one woman told the truth about her life?/The world would split open...." How would we narrate history in a way that includes the truth about our lives? To know this, some walls must come down, some doors open. One of the walls is The Wall of Transcendental Shame. Often, when we try to speak of our technological lives in a philosophical manner which includes experience, suffering, or exclusion, we are silently shamed – either within academia or within the swamps of convention. We are silenced, because people have been trained to turn away from intimate truths in the places where technoscience lives. Transcendental Shame is shame that has no real history, community, or moral order. The death-dealing words used to silence those who try to speak of this intimacy include "a bit confessional," "that's all very well, but where's she *going* with this?," "...and the point *is*...?"

Another wall is the Wall of Infinite Sequels. Such as "in future work we hope to extend this analysis to include such important issues as context, affect, and a more qualitative expansion of the independent variable, inequality." Or "It was beyond the scope of this study to include more variety in the sampling framework, such as women, minorities, or pay rates. Too much variability in the independent variables managed here would have produced a combinatorial explosion."

If we instead included the shaming and silencing processes as independent variables, and told the stories of combinatorial explosions, we would have a much richer idea of a history of ideas. Ideas occur between peoples, times, locations, and users of power – all of which have physical needs, ways of being attended to, and ways of being seen.

5. With respect to present and future inquiry, how can the most important philosophical problems concerning technology be identified and explored?

If we begin to refuse the types of walls alluded to above, we could stop using technology to sequester people and their experiences. If we begin with those who are excluded, shamed, and silenced, their lives will become the most important philosophical questions to be answered.

Resources

Millett, Kate. (1978 [1970]). *Sexual Politics.* NY: Ballantine.

Star, Susan Leigh. Forthcoming, 2007. "Living Grounded Theory: Cognitive and Emotional Forms of Pragmatism," In Bryant, Anthony and Kathy Charmaz, eds. *Handbook of Grounded Theory.* Thousand Oaks, CA: SAGE.

Bowker, Geoffrey and Susan Leigh Star. *Sorting Things Out: Classification and Its Consequences.* Cambridge, MA: MIT Press, 1999.

Star, Susan Leigh 1999. "The Ethnography of Infrastructure," *American Behavioral Scientist*, 43: 377–391.

Star, Susan Leigh and Anselm Strauss. 1999. "Layers of Silence, Arenas of Voice: The Ecology of Visible and Invisible Work", *Computer-Supported Cooperative Work: The Journal of Collaborative Computing*, 8: 9–30.

Star, Susan Leigh. 1998. "Experience: The Link between Science, Sociology of Science and Science Education," Pp. 127–146 in Shelley Goldman and James Greeno, Eds. *Thinking Practices.* Hillsdale, NJ: Lawrence Erlbaum.

Star, Susan Leigh and Karen Ruhleder, 1996. "Steps toward an Ecology of Infrastructure: Design and Access for Large Information Spaces," *Information Systems Research*, Volume 7:1, 111–134.

Star, Susan Leigh, 1992. "Craft vs. Commodity, Mess vs. Transcendence: How the Right Tool Became the Wrong One in the Case of Taxidermy and Natural History" Pp. 257–86 in Adele Clarke and Joan Fujimura, Eds. *The Right Tools for the Job: At Work in Twentieth Century Life Sciences.* Princeton: Princeton University Press, 1992. Translated into French, 1994.

Star, Susan Leigh. 1991. "Power, Technologies and the Phenomenology of Standards: On Being Allergic to Onions," Pp. 27–57 in A Sociology of Monsters? *Power, Technology and the Modern World*, John Law, Ed. Sociological Review Monograph. No. 38. Oxford: Basil Blackwell.

Hornstein, Gail and Susan Leigh Star, 1990. "Universality Biases: How Theories about Human Nature Succeed," *Philosophy of the Social Sciences*, Vol. 20: 421–36.

Star, Susan Leigh. *Regions of the Mind: Brain Research and the Quest for Scientific Certainty.* Stanford: Stanford University Press, 1989.

Star, Susan Leigh and James Griesemer. 1989. "Institutional Ecology, 'Translations,' and Boundary Objects: Amateurs and Professionals in Berkeley's Museum of Vertebrate Zoology, 1907–1939," *Social Studies of Science*, 19: 387–420 (1989). Reprinted in Mario Biagioli, Ed. 1998. *The Science Studies Reader*. Pp. 505–524. London: Routledge.

23

Isabelle Stengers

Professor, Chargée de cours

Université Libre de Bruxelles
Belgium

---

1. Why were you initially drawn to philosophical issues concerning technology?

My starting point was what may be conventionally called "the philosophy of theoretical physics." More precisely, it was associated with Ilya Prigogine's struggle to have irreversibility recognized as irreducible to any statistical approximation or coarse graining.

My first problem concerned what Prigogine had to face: the quasi-metaphysical authority of time-symmetric descriptions warranted observer-independent, complete description. In order to engage with this perspective, I needed to refrain from accepting the theoretical edifice of physics as value-neutral. Such an edifice is committed to what Kant called "practical questions," with the relevant difference that "humans" are surreptitiously replaced with "physicists".

I also refrained from reducing the questions "what is physics," "what must physicists do if they are to remain loyal to physics," and "what can physicists hope to reach" to answers that would merely reflect physicists' values. I could not adopt such a stance because of Prigogine's relentless and obstinate effort to "undo" the judgment against irreversibility. In order to be successful, what Prigogine had to produce was not just another interpretation of physics. Rather, he needed to generate a proposition that could redesign, in a consistent and precise way, what is considered to be the most glorious achievement of theoretical physics. In other words, what Prigogine had to produce are new practical insights, but in a demanding pragmatic sense: he had to propose new ways to intervene and measure, with the eventual emergence of new "facts".

This entanglement between two meanings of "practical," one pointing towards "commitment" (what I later called "obligation"), the other towards the way physicists deal with their "world" (i.e., the way they make it matter, or "work") has been at the source of what I have come to generically characterize as a "practice". In this sense, "generic" should not be confused with "general". The very point of a generic characterization is to define common questions the interest of which is the divergence of the answers they will be given. Every practice, be it scientific, technical, political, or ritual, may be associated with a generic "it works!" i.e., with an achievement. However practices radically diverge about the definition of what counts as an achievement; they are bound by diverging obligations.

Obligations, should not be reduced to a matter of following fixed rules, or obeying norms. Rules and norms ask for conformity while obligations are rather matters of concern and hesitation: they do not demand submission but imply the question of their eventual betrayal. Superstring theory, for instance, currently prompts the question, "Is this still physics?"

If practitioners are gathered not by fixed rules, which they can defend, but by obligations that are a matter of hesitation, vulnerability is another common feature of practices. When practitioners claim autonomy, no matter what doubtful means they use, we may have to understand it as expressing a vital need that the question of their obligations, and the hesitation this question entails be their own affair. A practice may be destroyed if obligations lose their power to gather practitioners around the question of what counts as an achievement.

My starting point thus led me to question any generality, be it the identification of physics with a model of rational knowledge, or any depiction that reduces a practice to either an instrumental or normative enterprise. It goes without saying that my reservations about fixed criteria are not neutral. It entails serious doubts about sciences that mimic experimental objectivity, without the obligations that relate experimental objectivity and achievement. On the other hand, in *La Vierge et le neutrino*, I described as a practice collective pilgrimages, with pilgrims traveling together towards a holy site where they hope to experience the Virgin's presence.[1] I claimed that these pilgrimages cannot be understood if the pilgrims are characterized as naively "believing" in supersti-

[1] Paris: *Les Empêcheurs de penser en rond*, 2006.

tions. Instead, the example illustrates the radical divergence between pilgrims' obligations and experimental practitioners' ones. In order to resist hierarchically aligning such diverging practices, we need to be able to affirm that both the Virgin and the experimenters' neutrino "exist". The mode of existence of the former has however nothing to do with the mode of existence of the latter because achieving relations with each of them implies diverging obligations. Recognizing this divergence opens us to an ontology that is pragmatic and irreducibly plural.

My "practice" approach thus leads me to distinguish between scientific and technical practices. Technical practices produce new modes of existence, the achievement of which demands taking into account many disparate dimensions (from gods to patents) and granting full importance to the question of *how* to interrelate all of the relevant actors. This characterization includes, in the same category, practices that are usually conceived of in oppositional terms: so-called traditional technical practices, and those which present and explain themselves as derived from a scientific definition of what they deal with.

By following this approach, I encountered the question of "technology". And this very approach led me to refuse to endorse the general definition of technology, i.e., "the application of science, especially to industrial or commercial objectives". The novelty designated by the term "technology" is irreducible to a matter of "application", and it may be described from many standpoints. Mine emphasizes non-locality and non-specificity. It depicts technology as an expanding process that transforms directly or indirectly everything that it succeeds mobilizing. Technology thus conceived does not have a privileged relation with scientific practices or with technical practices. Rather, it creates a new network environment, one that mobilizes sciences and techniques among other active ingredients.

I would propose, therefore, that technology is not a practice but rather something that "happens" to practitioners. It binds them. It even commits them to destroying their own practice. This is obvious, for instance, with information or computing practitioners. What I call "technology" demands that practitioners work at their own demise—at producing procedures that make possible a chain of command to be faithfully executed.

The strong and obvious partiality of this characterization of technology is deliberate. It is primarily meant to emphasize that we are not able to give a neutral definition of the transition leading from technical practices and achievements to technology because this transition is inextricably linked to what Marxists call "capitalism". Nevertheless, the characterization I propose is post-Marxist. It refuses the idea that capitalism, through the liberation of productive forces, could pave the way for its own overcoming. My view is compatible with Felix Guattari's essay in political ecology. There Guattari describes the triple ecological devastation: of the planet, of the collective capacity to imagine and create, and of the individual capacity to think and feel.[2]

2. What does your work reveal about technology that other academics, citizens, or engineers typically fail to appreciate?

On the issue of technology, I would present myself as a daughter to what I would not hesitate to name the "GMO event" in Europe. This name implies both unpredictability—it failed to happen in the U.S.—and the capacities of an event to transform present perspectives, both on the past and on the future. My standpoint is that of a witness, as it cannot be separated from the collective and transformative learning trajectory that made the GMO contestation an event.

The GMO event started with health concerns, then a sensitive issue in Europe. Sanitary crises such as the outbreak of mad cow disease had just happened. Also, the precautionary principle provided a tool for contestation but its scope was mostly restricted to health problems.

It was however only a starting point. If it is possible to speak about an event, it is because the tentative answers that public authorities provided not only failed to convince but rather helped extending the debate and opposition. It progressively became a matter of public knowledge that significant unanswered questions remained, and that reliability of expertise needed to be questioned. The debates also produced a reframing of the GMOs themselves. Today they are no longer deemed worthwhile innovations whose risks should be accepted despite some possible problems. Instead, they are viewed as vehicles for Intellectual Property Rights (IPR), the synonym of the capitalist expropriation strategy and the unsustainable development that follows.

---

[2] F. Guattari, *Les trois écologies*, Paris: Galilée, 1989. See also P. Pignarre and I. Stengers, *La sorcellerie capitaliste*, Paris: La Découverte, 2005.

What has been publicly defeated, therefore, is the motto, "Science produces technological innovation that answers human needs and solves human problems". When the GMOs failed to convince those who should have gratefully accepted them, when even the great information campaign launched by the Blair administration turned into a tool for the intensification of skepticism, the whole array of allies that this innovation aligned became perceptible. It is as if the unquestioned necessity had acted as a kind of Leibnizian vinculum, one that produced and was also being produced by, the empowerment of some protagonists and the silencing of others.

The initial absence of field biologists (an empirical, not experimental, science!) became conspicuous. Such absence made the internal hierarchy of sciences, and the correlative vast domain of empirical unknowns, perceivable. While lawyers and public authorities still appear to define the defense and promotion of IPR as a sacred duty, and while the WTO rules imply the necessity of final acceptance of GMOs, many now see such positions as active participants in the technology itself. What was presented as a progress is now felt as violence imposed by an occupying power.

While activists try and keep Pandora's box open, that is, work at increasing the scope and efficacy of all the objections and denunciations the silencing of which was taken for granted by past technological developments, European authorities now officially associate their conception of "governance" with the urgent need to avoid the repetition of such an event, with nanotechnology for instance. They proclaim their will to associate "citizens" to public assessment procedures in order to "reconcile the European public with its science". But this proclaimed will clashes with what is not only a familiar sociological standpoint, it is also a crucial part of the vinculum, namely the association of the problem of the public with a question of (mis)perception. Public questions, if empowered, will destroy any fixed frame. They will not accept the usual divide between what is to be decided by sound science and what is left for ethical debates. Instead, they will question the very fabric of the scene they are invited to assess, and demand accountable protagonists.

What academics, including many critical ones, engineers, and citizens typically fail to appreciate—but which the GMO event, with its unknown consequences, made perceptible—is that the casting that assigns their respective role to the aligned promoters of a technological development also deprives them of the capacity to account for this role. Not only are the GMOs not the crowning application of biology (but rather result from a rather poor molecular biology, with inflated ambition, i.e., "wait for the second generation!"), but they herald the direct mobilization of biologists through the so-called economy of knowledge, that is, an economy that also destroys what gathers them as practitioners. Already the IPR are a constitutive part of experimental biology, with the power to stabilize the gene-oriented framing of biological research. The point is not that scientific research is in the process of being dissolved into engineering. Engineering itself is being redefined, because technical achievement is not what matters for technology: a technical innovation can be the means for something else entirely.

We will have to learn how to interact with specialists whose obligations are dissolved and who are now primordially attached to partners imposing their own priorities—priorities that are indifferent to what forces practitioners to think, to hesitate, and also, eventually, to learn from outside objections and counter-propositions.

3. What, if any, practical and/or social-political obligations follow from studying technology from a philosophical perspective?

I would first of all assert that sophisticated historical studies are very important, but that we need other, non-deconstructive, narratives, as Donna Haraway would put it. It is not sufficient, for instance, to contest or deconstruct the linear connection between scientific achievements and technical ones. Nor is it sufficient to give the lie to the scientists defensive motto, "Do not kill the golden eggs goose, but feed us without imposing your questions and priorities."

Indeed, we know very well that the so-called geese were in fact actively collaborating with the production of the environment where their eggs would be golden. But today's narratives have to take into account that the goose's plea and argument do not work any longer, that the former respected gooses are now either happy players in the new setting or else in full disarray and therefore bound to angrily denounce those who would describe in terms of lies and privileges, i.e., what they experience as the destruction of their practice.

I would plead that critical (academic) narratives must learn how to address protagonists already in disarray. In particular, they should resist the temptation of the "we told you so," or the "well deserved!" The temptation to go along with the destruction of something we deem lacking with regards to our own criteria is exactly what the destroying process needs in order to proceed to case-by-case destructions that encounter minimal resistance.

It is easy to deride practitioners who complain about the "mounting irrationality" of an epoch that no longer honors disinterested knowledge. But I deem it important that their voices nevertheless be heard and relayed. One of the reasons why I forged the notion of "obligations" was to resist confusing the disarray of scientists with a refusal to renounce comfortable habits and privileges, as the promoters of flexible mobilization would invite us to do.

I would claim that it may be a social-political obligation to try to connect with scientists through tales that would prepare them not to parade as the defenders of rationality, but to make public in non-modern terms what matters for them, what makes them think and work, and the way their practice is now in the process of being destroyed. Such a social political obligation does not only concern scientific practices. All practices are threatened, and each time they appeal to modern justifications, they act as their own worst enemies. Indeed those justifications allow no room to obligations and thus invite to equate them with habits that can be freely modified following the demands of the moment.

These considerations also require us to resist the usual divide between so-called "traditional" techniques and "scientific" techniques. In other words, we need to resist something that has the power of what Deleuze and Guattari, in *A Thousand Plateaus*, call a "*mot d'ordre*". This power is the power to align, at one sweep, and before any possibility of hesitation, such disparate techniques as metallurgy and psychoanalysis, as both appear as purified from "beliefs". The point is not the validity of the theory of the unconscious or of metals, but the fact those modern techniques define themselves as deriving from a knowledge of what they "really" address, this addressed reality being sufficient to explain not only their own eventual success, but also the success of their non-modern predecessors ("they did not know it, but ... ") . In so doing, they may present their claim in universal terms, in contrast with particular local beliefs that were ingredients of so-called traditional techniques. Any philosophical perspective that ratifies this divide will be dominated by it. Its own questions will be organized around ready-made oppositions, such as "objective" and "subjective," "relative" or "universal".

All of these considerations matter when studying technology, that is, the aligning of techniques into networks that are indifferent to technical achievements. Indeed, what the processes of technological aligning needs is precisely the purification narrative: The technological networking of techniques needs techniques that present themselves as modern, purified from what never really mattered, from what can be by-passed, ignored or reduced to subjective, pliable resistances. What technological networking requires, then, is completing the job by defining the obligations of practitioners as also being subjective and pliable.

The tales we need would thus actively resist the "modern" dismembering of a technical achievement into a knowledge definition, with a claim to universality, and historical-cultural (including socio-economic and legal) contingent features that would shape any particular technique. This does not mean that we should celebrate such achievements, i.e., defend beneficent or benevolent techniques against their technological enslavement. Achievement does not designate something "good," but rather something coming into existence through a positive composition of multiple, heterogeneous ingredients. Again, it is a question of constructing a generic characterization of technical achievement, without any *a priori* differentiation between gods, scientific definitions and proofs, rites, patents, or the market. This characterization demands that every technical achievement be equally characterized as the grasping together of everything that it requires, mobilizes, or neglects, everything that it takes for granted, takes into scrupulous account, speculates upon, or defines as threatening. From this perspective, the modern "purification" narrative would itself be an ingredient. It does not define or separate; it signals a new distribution of what must be taken into account and what can be neglected.

Such a generic characterization is an active proposition. It opens up the possibility for practitioners to reclaim a practice that does not need purification—a practice the specific grandeur of which is the production of a togetherness actively inventing how to take into account what matters for the environment that will be transformed by this production.

4. If the history of ideas were to be narrated in such a way as to emphasize technological issues, how would that narrative differ from traditional accounts?

Emphasizing technological issues has something redundant about it. It is also a bit annoying from the standpoint that makes me think (not the only possible one, for sure) since the very issue is then technology itself, as it needs to be distinguished from technical issues. From this standpoint, it may well be that the very category of history of ideas must be questioned; the association between history and ideas very easily leads to still another version of the Great Divide between the "us" and the "others". To "have an idea" is certainly something rather commonplace and this experience probably also extends to apes, and other animals. But to have an idea and try to make it part of a history, designates something else entirely—what I would call an entrepreneurial culture, a culture in which ideas get linked to a history of innovation and in which ideas are explored as opportunities for actualization. Even if such culture may be said to have started in Greek antiquity, when specialists had to impress their wealthy contemporaries about the interest and novelty of their ideas in order to get reputation and reward, it is primordially "ours"; this includes the readers of this essay.

In *Adventures of Ideas*, Alfred North Whitehead emphasizes that the great Platonic invention is the definition of humans as "sensitive to ideas". Ideas are not produced by humans; rather the receptivity to the erotic power of ideas characterizes humans as such. Such a proposition may reverse some questions. Instead of asking about those ideas that kept so many cultures from recognizing their common belonging to humanity, we should perhaps accept that those who called themselves humans were produced by the power of what they called idea.

As Whitehead himself remarks, we should not exaggerate and equate Greece with the love of ideas: "*Even if you take a tiny oasis of peculiar excellence, the type of modern man who would have most chance of happiness in ancient Greece at its best period is probably (as now) an average professional heavy-weight boxer, and not an average Greek scholar from Oxford or Germany. Indeed the main use of the Oxford scholar would have been his capability of writing an ode in glorification of the boxer.*"[3] Nevertheless, the proposition that the history of ideas would address ideas as powers and not as human productions may be a fruitful one; it envisages ideas as other cultures may envisage ancestors, divinities or spirits.

It is possible to speculate that such a proposition would interest other cultures. The tale to be told about us would be a tale

[3] *Science and the Modern World*, New York: The Free Press, 1967, p. 204

about recklessness, about not honouring or cultivating the needed distance with what makes us humans. This is common wisdom indeed that powers that are not "fed" with due care and regard may turn into devouring ones. They can feed upon the reckless people who try and appropriate or instrumentalize them.

A history of the way those who called themselves humans came to see ideas as authored, or to qualify them in terms of truth or falsity, should proceed with caution however. If I am interested in "practical obligations," it is because those obligations do not gather "humans" as such. Practitioners may well present themselves in a modern environment as "having ideas," and use the modern public language that relates ideas with arguments and facts that should make them equally agreeable for every rational being; but they know that such an agreement has nothing to do with the questions and hesitations that make them practitioners. The contrast Bruno Latour emphasizes between "science in the making" and "science made" is not a deconstructivist one; it does not debunk shameful tricks behind the façade of anonymous rationality. Rather, it is Latour's expression of what makes the difference between practitioners and the *post hoc* fictions of humans as beings who test ideas without in turn being tested by them.

Because this fiction aligns and progressively destroys practices, it may well be the issue to be associated with "technology". The irony is that in this case, the very category of a history of ideas stammers. What are the ideas at work in nanotechnology, biotechnology, or information technology? We know for instance that nanotechnologies are predicted to produce a convergence transforming and correlating domains such as cognitive sciences, robotics, informatics, artificial intelligence and biology. What is striking is that the eventuality of such a convergence is independent of any idea that each of those domains would have for its responsibility to test and assess. The prediction is about the domains themselves and their irreversible invasion and transformation by an irresistible imperative. A nameless and devouring god arises.

5. With respect to present and future inquiry, how can the most important philosophical problems concerning technology be identified and explored?

The only philosophical problem concerning technology, as I address it, is that of learning which kind of framing of our common history weakens resistance, turning technology into some kind of logical development or fate. I would state that those framings are legion, and that they are rather systematically produced by the so-called academic world. It is here rather indifferent that this fate be denounced or identified as the triumph of rationality. The point is that those frames usually ratify what technology requires and produces the disempowerment of what it aligns. Debunking scientific practices, demonstrating how they always served what is now destroying them is part of the game. It is an easy game and also an academically friendly game. Academic palms crown those who successfully present themselves as more lucid than their colleagues who "would still believe that ... "

The question, here, does not concern belief. I do not "believe," for instance, that the European event of collective recalcitrance against GMOs has, as such, the power to oppose the rule of IPR over agriculture. I would rather state that the concerned groups' process of empowerment and their reclaiming actions are one of the not so many unknowns in our future. Those groups potentially include scientists resisting despair and engineers feeling dishonored. The philosophical problem would then be to track down all philosophical positions that entail an *a priori* distrust against such processes and actions.

I am not only referring to the political question of minority groups who do not consent to classical mobilization, but also to the empowerment techniques that some of them experiment. When the neo-pagan witch Starhawk writes about magic as an art or a craft, and about the collective rituals invoking the goddess as experimental empowering techniques, she is bound to cause a very well-entrenched distrust against manipulative techniques that play on human weakness and suggestibility. She goes against the grain of the humanistic tradition that celebrates human emancipation—what Kant described as the coming into maturity of humanity at last able to take its fate into its own (moral and critical) hands. I would propose that the academics who feel uneasy about what those witches' techniques, which fear some kind of irrational regression, are humanist in this sense. And I would also propose that their reaction manifests the solidarity between this tradition and the modern dismembering of techniques through the use of blanket categories ("suggestion," "superstition," "beliefs") designating what we should free ourselves from.

The rather sorry state of the academic world, pathetically unable to resist submission to the blind assessment technology targeting its producers and productions, offers testimony enough that critical thought is by itself unable to empower collective innovative reclaiming processes. I would propose that we try to experience what it takes to define ourselves as in urgent need to learn from those who experiment what it is so easy to criticize and even deride. The kind of reaction we would then have to face may well confirm Starhawk formula that "the smoke of the burning Witches still hangs" in the (academic) air we breath.

Bibliography

Last books in French

Penser avec Whitehead. Une libre et sauvage création de concepts, coll. " L'ordre philosophique ", Paris, Le Seuil, 2002.

L'hypnose entre science et magie, Paris, Les empêcheurs de penser en rond, 2002.

With Bernadette Bensaude-Vincent, *100 mots pour commencer à penser les sciences*, Paris, Les Empêcheurs de penser en rond, 2003.

With Philippe Pignarre, *La Sorcellerie capitaliste. Pratiques de désenvoûtement*, Paris, La Découverte, 2005.

La Vierge et le neutrino. Les scientifiques dans la tourmente, Paris, Les Empêcheurs de penser en rond, 2006.

Books translated in English

With Ilya Prigogine, *Order out of Chaos*, New York, Bantam Books, 1984.

With Léon Chertok, *A Critique of Psychoanalytic Reason*, Stanford, Stanford UP, 1992.

With Bernadette Bensaude-Vincent, *A History of Chemistry*, Cambridge Mass., Harvard University Press, 1996.

Power and Invention. Situating Science, Minneapolis, University of Minnesota Press, 1997.

The Invention of Modern Science, University of Minnesota Press, 2000.

24

Lucy Suchman

Professor, Anthropology of Science and Technology

Centre for Science Studies/Department of Sociology
Lancaster University, UK

---

1. Why were you initially drawn to philosophical issues concerning technology?

Like many of us, my engagement with technology studies emerged as an unexpected coincidence of intellectual interests, political anxieties and personal circumstances. As an entering student at the University of California at Berkeley in 1968, I found my way into the field of American anthropology under the sign of 'studying up'. This was an initiative that called for a redirection of anthropological attention from those who had been the field's historical research subjects – in brief, those people who were the objects of colonial administration – to those who occupy positions of power and privilege within dominant institutions. This translated for me into an interest in the United States, and more specifically U.S. corporations, which led me through a series of fortuitous circumstances to Xerox's Palo Alto Research Center (PARC). PARC led me in turn to questions of technology, and the latter has been the frame through which I've worked ever since.

On my arrival at PARC I took up a position as a Research Intern in a group of computer scientists engaged in the early days of what were then called office information systems. It was clear that they believed that procedural office work was ready-made for automation, because computers carry out procedural instructions just as, they assumed, office workers do. In contrast, my own immersion at the time in the ethnomethodological and conversation analytic literature suggested that what it means to conduct work according to procedure in an office is something quite different. So I imagined that I might take as the focus of my dissertation research the most apparently routine and procedural form of office work I could find, and then explore the question of whether or not the assumptions that were being made at the time about office work, which were informing the design of information systems, were sound ones.

As a pilot for that larger project, I did a study in the accounting office at PARC. I sat with people while they processed expense reports and accounts, and developed an analysis of what that work specifically entailed. I drew here again on ethnomethodological arguments regarding the relation between procedures, instructions, plans – any kind of schematic and prescriptive representations of how things should be done – and the practical work of carrying them out. I observed that it wasn't the case that people followed the procedure most of the time and then sometimes deviated from it. Rather, in order to follow the procedure you had to engage in continuous forms of ad hoc, and often creative, reasoning and action. Members of the accounting office were oriented to the fact that auditors might come to the accounting office, open any file drawers at random, pull out a file and inspect it. Their task, accordingly, was to create files that, if read by an informed person, could be read as the record of something that was done according to procedure. This meant establishing an accountable relationship between procedural instructions and specific cases, such that the work became, arguably and demonstrably, action in accord with the rules.

In this context I started to think about the procedures themselves as artifacts, produced in the work's course and taken as resources for doing it. That implies, in turn, a very different kind of computer support. This argument appeared to resonate with discussions emerging at the time in the area of what came to be called "knowledge-based" systems. The idea was that research and development needed to shift from automation to developing systems designed as tools to be used by knowledgeable people or human experts. I tried to engage with that line of discussion, but on the premise that if you look at any form of human work closely enough, you discover that it's knowledge work. The trick in designing information systems, then, is to introduce bits of automation that will fit in to the work and do useful things, while leaving the discretionary space required in order to get the work done.

Through these studies I became increasingly identified as a kind of constructive antagonist within the Artificial Intelligence (AI) community, particularly among those involved with the project of designing so-called intelligent, interactive machines. This led to the study that became my dissertation (recently republished within an expanded discussion of developments both in science and technology studies and AI, under the title *Human-Machine Reconfigurations*). This was, again, a close examination of the assumptions about action and communication that were operating within the AI community at the time, and that were the basis for expert systems and intelligent interfaces. I was trying to question those assumptions and work out an alternative, based on a similar argument about the relationship between prescriptive representations and the contingent work of enacting them in practice.

2. What does your work reveal about technology that other academics, citizens, or engineers typically fail to appreciate?

I need to begin by questioning some of the assumptions implicit in this question, and hope that may serve as a kind of an answer.

First, it could never be the case that the work of any person would be so unique that it would offer insights that no one else shared – if that were so, it would clearly be unintelligible! I live within a nexus of multiple, partially overlapping research networks including anthropology, science and technology studies (STS), feminist research, and computing, all of which infuse my work. Second, the implication that the process involved is one of revelation suggests that technology studies are about the discovery of already-existing realities. But one of the major insights of science and technology studies is that subjects and objects, what we take to be knowledge and that which we constitute as the real, emerge together out of sociomaterial practices. This applies to technological practices as well as to our activities as STS researchers. I suppose that it is precisely this conceptualization of technology and of its study that is distinguishing, not of my work alone but of the community of scholarship to which I hope to belong and to contribute.

With that said, the understandings of technology that resonate for me are those that return an artifact to the arrangements of which it's part and through which it takes its significance. And the perspective that I have to work with that seems most distinctive comes from the multiplicity of locations that I've been able to inhabit within the worlds of technology research and development. That gives me a deep sense for the what Harold Garfinkel names the "demonically wild contingencies" that characterize the process of making and using artifacts, particularly insofar as the sites of professional design and ongoing design-in-use are increasingly distant from one another. I agree with Phil Agre that the parochialism of professional design is entrenched in profound and multiple ways in the organization of contemporary technical practices (1995: 77). My work has been concerned with some of the ways in which established arrangements of product development systematically separate professional designers from prospective technology users, and the range of proxy figures that have been devised to fill the gap. Experimental subjects are taken to speak on behalf of people encountering technologies in their everyday lives, and human subjects are increasingly replaced, or at least augmented, by scenarios and personae – synthetic and imaginary use settings and technology users drawn from more and less extensive encounters with indicative persons and sites. These stories and characters stand in for more distant and unruly sites and subjects, and make design manageable within the "time and money" constraints of an increasingly intensified, investor-driven market for new products and services.

At the same time, I've questioned just how specific and determining what Steve Woolgar (1991) has named the designer's "configuration of the user" is, in either design imaginaries or specific situations of use. It's increasingly clear to me that there is no stable designer/user "point of view," nor are imaginaries of the user or settings of use inscribed in anything like a complete or coherent form in the object. To grasp the design/use relation I believe that we need to see the designer's view of the user as at once more specific and less. More in that it's specifically located within the various sites, imaginaries, exigencies and practices that comprise professional design, and less in that artifacts are characterized by greater open-endedness and indeterminacy with respect to the question of how they might be incorporated into use.

3. What, if any, practical and/or social-political obligations follow from studying technology from a philosophical perspective?

If I translate the term "philosophical" here to "anthropological," this question becomes more answerable for me. As I've indicated in response to the previous questions, my twenty year engagement with worlds of research and development gives me a strong sense of the indeterminacies of technology design, and the enormous reliance of professional designers on the generative work of technology users in making sense of the artifacts that they encounter and incorporating them into their lives. In the late 1980s, I and a small number of colleagues succeeded in establishing a research group at Xerox PARC dedicated to anthropologically-based studies of work and technology on the one hand, and cooperative forms of system design on the other. We increasingly saw these two agendas as related, insofar as our studies and our design practices were aimed at reconstructing what knowledges and interests are taken to be relevant to the development of new technologies. The project was to open up the boundaries of legitimate knowledges and interests from the research labs and development organizations to other sites.

One aspect of our intervention was to argue for the value of specific, heterogeneous and non-reductionist representations of sites of work and practices of technologies-in-use. The context for this was the mode of systems development that was, and largely still is, prevalent, which takes the form of attempts to create generic representations of user requirements. The designers in Xerox, including our friends in the product organizations, worked under tremendous pressure to deliver user requirements in the form of bullet points that could be translated directly into design features. So ideally you'd have a user requirement that would tell you exactly what you should do in terms of designing (an ideal invariably disappointed in practice). User requirements were typically generated through focus groups, if a technology was in an initial stage, or usability testing if there was a system under development. These were, in effect, laboratory-based methods where people would be brought into a contained environment, typically paid, and asked to engage in these kinds of exercises. There was a major program within Xerox, for example, called 'Voice of the Customer,' where transcriptions from focus groups would literally be cut into sound bites, sorted, and grouped into categorical "concepts" that would be translated as requirements. This assumes that what you're eliciting from users are ideas, packaged as utterances, though in this case the "idea" is even decontextualized from its utterance, let alone from any wider activity that it might be part of. And what you get are desiderata like "easy to use," those phrases that then circulate around but what they actually mean, or what it means to translate that into design, is far from clear. With the result that designers are left where they started, to make it up.

So what we were doing instead was to negotiate access to organizations where we might work with people identified as prospective users of technologies under development. We would make a very specific connection within an organization and then enter into a fairly long term, very intensive relationship combining ethnography and co-design. The product of that engagement would be rich corpora of videotapes, images, stories, moments and events, as well as a prototype technology configuration. Our idea was that these materials should be a resource to which questions could be brought, and from which further questions could be generated, that could then be taken back. We had a fantasy that there might be long term relationships like this, so that you would have a partner with whom, on an ongoing basis, you'd be engaged in research. This was in part our response to the claim that "we don't have time for such engagements, product development works on shorter time cycles, we don't have time to be going out doing ethnographies." Our response to that was, the model here is not that you sit around on your hands while we go out and do our ethnography, then we bring back our 'findings' and hand them over to you, and you smite your foreheads and say "Oh, now we understand" and then design the system. Rather these things should be running in parallel, and we should be connecting continually along the way so there's a generative interaction between them.

This way of working meant that we were very resistant to producing "requirements" out of these materials, the things that we were learning through these engagements. We instead worked to come up with new research objects; for example, we talked about what we called 'working document collections' as an important focus for design. These are the documents in between those on your desk and those in the archive – the filing cabinet being the obvious precedent. So we did work to effect those kinds of translations. More generally, we tried to combine forms of resistance to business as usual with attempts to open up the practices of design so that those who are its subjects/objects could be more present. And in order for us to tell a credible story of how design might be otherwise, it was important to us to enact the practice that we were recommending – rather than say to designers "you should go out, spend a bunch of time with users" we wanted to create these projects that were pilot, small scale explorations of what we were arguing for, so that we would know ourselves what we were talking about.

In a Ph.D. seminar at University of Toronto where we discussed these things recently, Joseph Ferenbok suggested that we might think of this as another sense of "sustainable design," not only in terms of the 'life cycle' of the product, but of the creation of sustainable design processes, based on sustained relationships between technology designers and users. This insight is relevant as well to the perennial question of how more participatory design practices make a difference to the resulting product. It's crucial to understand that it's not simply a matter of a change to the features of the artifact, but rather that the labors of its production have been different, those who might make use of the artifact have been in relation with it, have been engaged in processes of incorporation and familiarization, fitting the artifact to a practice, from the beginning. So if obligations are to make a difference, the premise is, you can't address the artifact in itself, you've got to make differences in the processes and relationships through which the artifact comes into being.

4. If the history of ideas were to be narrated in such a way as to emphasize technological issues, how would that narrative differ from traditional accounts?

I have no idea! Seriously, it seems to me that one of the great contributions of science and technology studies has been to replace the idea of a history of ideas with investigations of the specific cultural, political/economic and practical arrangements that comprise "thinking" at a particular historical moment. Most famously, perhaps, Steve Shapin and Simon Schaffer have traced the enlightenment legacy of contemporary technoscience through its materializations in the natural philosophy of the 17th century, but many other contemporary social historians have contributed as well (think of Geoffrey Bowker and Paul Edwards on cybernetics, as well as Bowker's work on memory practices). Jessica Riskin has edited a fabulous collection titled *The Sistine Gap*, which explores the traffic over the past several hundred years in European thought between imaginaries of human and machines, the natural and the artificial, life and mechanism. She emphasizes that throughout this history neither human nor machine provide a fixed point of reference for the other, but rather the traffic goes both ways.

Others like Karen Barad, Donna Haraway, John Law, Annemarie Mol and Helen Verran have urged a mixing of ontics and epistemics, emphasizing the inextricable relations between constituting and knowing, objects and subjects. The ethnomethodological approach to this question has involved what Mike Lynch sums as "respecifying the central topics in epistemology by identifying them as commonplace discursive and practical activities" (1993: 5, fn. 9). Karin Knorr Cetina has elaborated a notion of 'reconfiguration,' which for her is a way of thinking about laboratories as arranging scientists, instruments, objects and practices so as to construct nature in particular ways (Joan Fujimura's work follows this line of inquiry as well). The anthropologist Tim Ingold observes that twentieth century anthropologists and prehistorians posit a juncture that effects the 'beginning of history', repositioning technology from a functional attachment of the organism to an element of a cultural system, uncoupled from biological evolution. With this "breakthrough" into culture, the story goes, technological change took off. Ingold's way out of the nature/culture divide inherent in this story is to expand the frame, from a posited genotypically determined design for individual organisms to "transformations in the whole field of relationships within which they come into being" (2000: 366). A narrative based in the assumption that tools have inherent and measurable properties gives one history, while another oriented to the wider frame of tools-in-use gives another. A social and historical arrangement that privileges the translation of complexity from human activity into the black box of the tool effects a different system of measurement of change than one that considers more extended sociomaterial assemblages. The basis for these comparisons lie in prior stories, in other words, regarding conceptions not only of change but of technology itself.

5. With respect to present and future inquiry, how can the most important philosophical problems concerning technology be identified and explored?

I don't believe that any of us is in a position to define what are the most important philosophical problems concerning technology (or that such problems exist in any timeless form). But I could offer some thoughts on what seem to me compelling directions for technology studies. Science and technology studies over the past two decades have changed the way that we conceptualize sameness and difference among and across animals, humans and machines, as well as the boundaries, extensions and enfoldings of human/nonhuman bodies, and the heterogeneity and distributions that comprise agency. There has been tremendously exciting work within STS looking at very specific sites and configurations of human/nonhuman agency in scientific knowledge practices (see for example Pickering 1995; Fujimura 2005). Natasha Myers (in press) documents how coming to know molecular structures involves their incorporation not simply as mental representations but bodily understandings; Rachel Prentice (2005) examines how in cases of surgical simulation and technology-mediated surgery the surgeon's body reunites what the technology seemingly makes more distant; and Natasha Schull (2005) investigates seductive entertainments where digital gambling machine developers and compulsive gamblers conspire to intensified forms of body/machine merging. This attention to questions of nonhuman agency has been a tremendously restorative antidote to exclusively human-centric social sciences.

But at the same time that this body of work documents the specificity and intimacy of human/nonhuman mixings, I think that it also puts us in a strong position to press further on questions of difference, moments of distinction that matter, both conceptually and practically, between animals (including humans) and machines. I'm thinking here, in particular, of the need to intervene in initiatives generated within the worlds of government-sponsored research and development. To pick just one example that came to my attention recently, the Office of Naval Research (ONR) is currently requesting proposals for research on "Peer-to-Peer Embedded Human Robot Interaction." The request notes, "one issue that limits the use of robotic and autonomous systems in urban environments is their inability to recognize and interact with persons who may be either non-combatants or threats." The solution to this is imagined as improved robot understanding of human goals, plans, activities and communications, based on research in machine vision and, specifically, "new capabilities in human activity and gesture recognition." Given that my own work over the past twenty years has been dedicated to close critical analysis of the bases for research on 'socially aware' machines, I'm tempted to allocate this Request to the realm of fantasy. But it doesn't take a close reading to see its links with favorite projects currently underway in U.S. research and development laboratories, and to project the flow of new funds into those sites. Moreover, the unfeasibility of inscribing the kinds of humanlike capacities that are called for into robot weapons systems means not that these efforts will be abandoned, but rather that they will of necessity fall back onto the worst kinds of profiling, too much of which we already see at work in the so called 'war on terror'. These developments call urgently for close and critical engagement by science and technology scholars. And this is a moment when human/machine differences matter, when the human capacities involved in the always perilous task of co-constructing intentions as friend or threat need to be elaborated and specified, rather than simplified and reduced to rule.

The Navy's call for improved "force protection" points to a broader direction for STS research as well. My colleagues Maggie Mort and Celia Roberts and I have begun to sketch out a research agenda focused on relations between action at a distance and bodies in proximity across the domains of military weaponry and healthcare technologies. The ONR call cited above is just one among many examples of the military's obsession with enhanced capabilities for action at a distance. Another colleague, Theo Vurdubakis, has begun to trace the many manifestations of this orientation to "protection" of the U.S. soldier, keeping her/his body in a safe place by projecting agency through technology, remote controlled and ideally autonomous. Theo suggests that we understand the suicide bomber as the poignant response to this investment: in the absence of high tech what's available is your own body, the ability to put your body in proximity with others and become, yourself, a weapon. At the same time, albeit to very different ends and driven perhaps more by discourses of economy and efficiency, research and development in healthcare technologies is increasingly oriented not only to telemedicine but robotics, and in particular to 'socially intelligent' robots, or robot companions. This is another area where it's really important to understand how specific human-machine configurations matter, what things can be done at a distance, and what difference it makes to have bodies in contact with each other. How, we might ask, do technological imaginaries circulate across these seemingly disparate, if not opposing, domains?

Acknowledgements: My thanks go to Andrew Clement and Terry Constantino, Diane DeChief, Joseph Ferenbock, Adam Fiser, Brenda McPhail and Karen Louise , Ph.D. students at the Faculty of Information Studies, University of Toronto for turning this into a truly "lively conversation".

Selected Bibliography

Agre, Philip (1995) Conceptions of the user in computer systems design. In P. Thomas (ed.), *The Social and Interactional Dimensions of Human-Computer Interfaces* pp. 67–106. Cambridge, UK and New York: Cambridge University Press.

Barad, Karen (2007) *Meeting the Universe Halfway: Quantum Physics and the Entanglement of Matter and Meaning.* Durham, North Carolina: Duke University Press.

Bowker, Geoffrey (1993) How to be Universal: Some cybernetic strategies, 1943–1970. *Social Studies of Science* 23: 107–127.

Bowker, Geoffrey (2005) *Memory practices in the sciences.* Cambridge, MA.: MIT Press.

Edwards, Paul (1996) *The Closed World: Computers and the Politics of Discourse in Cold War America.* Cambridge, MA: MIT Press.

Fujimura, Joan (2005) Postgenomic Futures: Translations Across the Machine-Nature Border in Systems Biology. *New Genetics & Society* 24: 195–225.

Garfinkel, Harold and Rawls, Anne (2002) *Ethnomethodology's program: working out Durkeim's aphonism.* Lanham, Md.: Rowman & Littlefield Publishers.

Haraway, Donna (1997) *Modest Female: Man Meets Oncomouse* (New York and London: Routledge, 1997).

Ingold, Tim (2000) *The Perception of the Environment: Essays in livelihood, dwelling and skill.* London and New York: Routledge.

Knorr Cetina, Karin (1999) *Epistemic cultures: how the sciences make knowledge.* Cambridge, Mass.: Harvard University Press.

Law, John and Mol, Annemarie (2002) *Complexities: Social Studies of Knowledge Practices.* Durham and London: Duke University Press.

Lynch, Michael (1993) *Scientific practice and ordinary action: ethnomethodology and social studies of science.* New York: Cambridge University Press.

Myers, Natasha (in press) Molecular Embodiments and the Body-work of Modeling in Protein Crystallography. *Social Studies of Science.*

Pickering, Andrew (1995) *The Mangle of Practice: Time, agency and science.* Chicago: University of Chicago Press.

Prentice, Rachel (2005) The Anatomy of a Surgical Simulation: The mutual articulation of bodies in and through the machine. *Social Studies of Science* 35(6): 837–866..

Riskin, Jessica (ed.) (2007) *The Sistine Gap: Essays in the history and philosophy of artificial life.* Chicago: University of Chicago.

Schull, Natasha (2005) Digital gambling: The coincidence of desire and design. *The ANNALS of the American Academy of Political and Social Science* 597: 65–81.

Shapin, Steven (1994) *A social history of truth: civility and science in seventeenth-century England.* Chicago: University of Chicago Press.

Shapin, Steven and Schaffer, Simon (1985) *Leviathan and the air-pump : Hobbes, Boyle, and the experimental life.* Princeton, N.J.: Princeton University Press.

Suchman, Lucy (2007) *Human-Machine Reconfigurations.* New York: Cambridge University Press.

Verran, Helen (2001) *Science and An African Logic.* Chicago: University of Chicago.

Woolgar, Steve (1991) Configuring the User: the case of usability trials. In J. Law (ed.), *A Sociology of Monsters: Essays on Power, Technology and Domination.* London: Routledge. *Dreaming the Dark*, Boston: Beacon Press, 1997, p. 219.

About the Editors

Jan-Kyrre Berg Olsen has a Ph.D. in Science Studies and teaches philosophy at Roskilde University. Besides *Five Questions in Philosophy of Technology* he is also editor of several volumes forthcoming later this year and in 2007, including *Technology and Science: Epistemological Paradigms and New Trends*, special issue of *Synthese*, Springer; *New Waves in Philosophy of Technology*, London: Ashgate Publishing; *A Companion to Philosophy of Technology*, Oxford: Blackwell Publishing Ltd.

Evan Selinger is an Assistant Professor of Philosophy at Rochester Institute of Technology. He has written many articles about issues raised in the philosophy of technology, philosophy of science, science and technology studies, and phenomenology, and has also edited or co-edited several books on these topics. These books include: *Chasing Technoscience: Matrix for Materiality, Postphenomenology: A Critical Companion to Ihde, The Philosophy of Expertise*, and *New Waves in Philosophy of Technology.* He is also the Book Review Editor for *Human Studies* and a member of the Group for Logic and Formal Semantics.

About Philosophy of Technology

Philosophy of Technology is a collection of short interviews based on 5 questions presented to some of the most influential and prominent scholars in this field. We hear their views on technology, its aim, scope, use, the future, and how their work fits in these respects.

> *Olsen and Selinger have given us a stunningly lucid presentation of the behind-the-scenes thinking that goes into the published work of twenty-four of the top philosophers of technology. Their book is a must-read for anyone interested in the present and prospects of our technological milieu.*
>
> —**Larry A. Hickman**
>
> Center for Dewey Studies
>
> Southern Illinois University Carbondale

> *As successive technological revolutions transform our everyday life, the philosophy of technology becomes more and more relevant to crucial issues of the contemporary moment. Philosophy of Technology: 5 Questions interrogates major philosophers of technology who address key issues in the field and lay out a wealth of positions. The result provides both insight into specific thinkers and an excellent overview of the field today in the philosophy of technology.*
>
> —**Douglas Kellner**
>
> George F. Kneller Chair in the Philosophy of Education
>
> UCLA

Technology is too important a topic to be left to the professional philosophers. Its neglect is a scandal. Philosophy of technology is at last undergoing a revival under the spur of contributors from science studies, anthropology and feminist scholarship. Many of the leading figures are interviewed for this book which provides insights into their thinking on technology, their goals, and their research strategies. The volume as a whole is an indispensable guide to how the modernist paradigm is slip sliding away—different road maps to amodernity and postmodernity are offered. There is even a welcome smattering of thoroughly modernist reactionaries.

—**Trevor Pinch**

Professor of Science & Technology Studies

Cornell University

Index

Printed in the United States
99608LV00001B/114/A